INTRODUCTION TO FOOD SCIENCE

A Kitchen-Based Workbook

AN OVERVIEW

SECOND EDITION

INTRODUCTION TO FOOD SCIENCE

A Kitchen-Based Workbook

AN OVERVIEW

SECOND EDITION

Dale W. Cox

AN **EDIBLE KNOWLEDGE®** SERIES WORKBOOK

ISBN: 978-1-948515-03-0
Ebook ISBN: 978-1-948515-04-7

Library of Congress Cataloging-in-Publication Data is available.

Published by Beakers & Bricks, LLC

Cover art painted by Jared T. Slack
Cover design by Glen M. Edelstein
Interior design by Glen M. Edelstein

Photographs by Dale W. Cox, unless otherwise specified in Pictures and Illustrations Attributions
Edible Knowledge® logo art by LeAnne Cox
Edible Knowledge® is a registered trademark of Beakers & Bricks, LLC
Printed in the United States of America

Beakers & Bricks, LLC
PO Box 1014
Asheboro, North Carolina 27204
www.beakersandbricks.com

This workbook is dedicated to my parents who encouraged me to write about my early acquisitions of food science knowledge. Special thanks to Dr. Oscar Pike for his help in reviewing the manuscript.

Disclaimer/caution: Experiments in this workbook may involve high temperatures or sharp knives. Please take every precaution to avoid injury.

CONTENTS

INTRODUCTION

Why do crisp cookies sometimes become soft? Why does fresh bread become hard if it's left out? Why can you store an unopened package of some types of snack meats at room temperature for months when refrigerated meats spoil when left out even for a few hours? Mold grows on some baked goods, but not others... What's up with that?! Why is some ice cream really hard and others soft when they come from the same freezer?

These are just a few of the many interesting questions that food science answers. During my first classes in college, I had many, many "Aha!" moments when I finally understood what I'd seen while preparing food.

Breakfast Cereal[1]

As a young person, I loved cooking, especially baking, and I often wondered about many of the transformations that happened:

- Why do pie crusts sometimes turn out nice and flaky, and other times tough?
- Why do egg whites turn white when they're cooked?
- Why does beef turn gray when it's fried?
- How come green vegetables like broccoli change from bright green to olive green when they're cooked?

I still remember the excitement I felt as I learned the answers to these questions, and always thought other scientifically curious people such as you might find them interesting. Through this workbook series, you'll explore some incredible science that explains what happens while food ingredients are transformed into a finished food, whether in your kitchen or in a food production facility. If you're curious, you'll likely feel a similar sense of wonder as I had when I was learning it myself. You'll use your kitchen as your laboratory and be able to watch concepts unfold right before your eyes!

The Edible Knowledge® workbook series is designed to be used in the fifth through twelfth grades as a science supplement, especially among those who are homeschooled. Assignments for further instruction are included for those interested, and many kitchen experiments are listed. Most experiments result in something that you can eat! It's even possible that through this introduction, you'll discover that you're interested and want to find out more. Note: Some of the questions and assignments deliberately require thinking beyond material that's presented in the workbooks, rather than a simple regurgitation exercise. These are challenging and require the use of some creativity, which is a requirement for scientists.

I hope that you find your introduction to food science as enjoyable as I did!

CHAPTER 1
FOOD SCIENCE BASICS

What is food science?

The study of the physical, biological, and chemical makeup of food and the concepts underlying food processing. Food technology is the application of food science to the selection, preservation, processing, packaging, distribution, and use of safe food."[i]

Food science incorporates most of the sciences, requiring training in microbiology, biology, chemistry, physics, engineering, and statistics, as well as specific classes associated with food chemistry and animal science, to name a few.

Books[2]

If you like science but don't like any one type enough to want to be a chemist or a microbiologist, for example, food science might be a good fit for you!

[i] Definition of food science quote: The Institute of Food Technologists, established in 1939 for food system professionals to collaborate and improve food production worldwide; members in 90+ countries. http://www.ift.org/about-us.aspx. Accessed 13 November 2018.

Using myself as an example, for the last couple of years in high school, I wanted to become a physician. But not everything turns out as planned! After working as an emergency medical technician for an ambulance company for a while, I knew that I loved working with patients. However, based on the advice of some dissatisfied doctors, I started looking around for something different. I loved the sciences and the orderliness involved in experimentation. However, I didn't love any one of them enough to want to do just that one thing. I didn't want to be a chemist, biologist, or microbiologist.

Eventually, a friend introduced me to food science. In this field, many science disciplines are incorporated in an applied manner almost every day. If you find that you also enjoy applied science, you may also enjoy a career in food science. Something to think about!

Food science isn't a well-known college major or profession. Most people don't know about it, so they never think of it as a career. In terms of job security, one of the last things that people do when money becomes scarce is stop eating! As a result, food science professionals are in high demand, resulting in jobs that tend to be fairly secure. In fact, even when many college graduates struggle to find jobs, such as during a recession, most food science students have secure employment before they leave school.

In the next section, you'll see some career options open to food scientists. If after going through this workbook you find yourself interested, a bonus chapter gives additional insight into the education required, including specific examples of course schedules from a couple of universities.

FOOD SCIENCE CAREER OPTIONS

Many other career paths are available to someone with a food science degree. Job openings exist in academia (at a university or college), in quality control, research and development, sensory science, flavor chemistry, and management. Below are some of these jobs described in more detail, with the salaries you can expect when working in those positions. Depending on the role, salaries in the food science industry can be quite high. You can make a good living as a food scientist, even without going into management.

Academics

Many people enjoy teaching others about what they love. You might as well! Opportunities exist to teach food science in high schools, but particularly in colleges and universities. Teaching a high school food science class might be done by a science teacher, and not necessarily a food scientist.

Graduation cap[3]

To teach at a university or college, generally a PhD (doctorate degree) is needed. At this stage you'd be a professor and likely be required to continue cutting-edge research in a very specific area of food science. College professors teach classes, possibly oversee students working on a graduate degree (a degree such as a master's or doctorate), and have undergraduate and graduate students helping in the research.

Teaching at a college with a PhD can bring in an annual salary averaging $100,000.[ii]

Sensory Sciences

The older generation might remember the Pepsi vs. Coca-Cola taste tests on television commercials. This was an example of sensory science that was used by a company to promote its products. Food companies, especially the larger ones,

A blind taste test[4]

[ii] Salary range reference: "2017 IFT Employment & Salary Survey Findings," p. 10. The Institute of Food Technologists.

employ sensory scientists to carry out consumer taste tests on products the company is developing or changing.

As an example, a fictitious company we'll call Scrumptious Discs makes a chocolate chip cookie. Their chocolate chip supplier had a fire at their facility and will be unable to supply chocolate chips for quite some time. Since chocolate chips are an important finished product component, the company doesn't want to annoy or alienate any of their loyal customers (aka *consumers* or *users*).

Research and Development are strategically important departments for any food company. They're frequently found together and abbreviated R&D.

Their R&D scientists locate a new supplier, oversee the production of finished product samples, and send them for sensory testing. Sensory scientists work with current cookie consumers to test the chocolate chips in the finished product and prove they're the same as the old ones.

To work as a sensory scientist, you need a bachelor's degree in food science with an emphasis in sensory science, or a master's degree in the sensory sciences. You also need a good understanding of statistics and know how to properly design questions and present food samples to avoid invalidating the results. It's a tricky business!

For example, "first order bias" describes the fact that whatever you taste first seems more flavorful than whatever comes after, even if they're the same. For this reason, among others, the samples are presented randomly to consumers to taste. Otherwise, the test results might indicate that the chocolate chips are the same when in fact they aren't.

Salaries for sensory evaluation specialists average $76,000 per year.[iii] Sensory scientists typically work in a main research facility, but they may oversee work done by quality assurance/ quality scientists in production facilities.

Food Engineering

Engineering is the application of knowledge, such as mathematics, science, and observed data, to the physical world bdesigning and making machines and processes to accomplish tasks. Food engineers apply food science in the physical world. Some universities offer degrees in food engineering, with stud-

Engineer at work[5]

[iii] Salary range reference: Kuhn, Mary Ellen, and Margaret Malochleb. "2017 IFT Employment & Salary Survey Report." The Institute of Food Technologists. Food Technology, p. 30, March 2018.

ies heavy in food science but also including more engineering principles. Some chemical engineers get a master's degree in food science. Many employers look hard for professionals with this desirable credential combination.

Examples of jobs for food engineers include working on an R&D team designing a process for a new product, improving the design of existing equipment from a maintenance or efficiency standpoint, or as a consultant who visits food plants and helps troubleshoot processing problems.

Food engineers can expect annual salaries averaging $108,500 per year.[iv] Their jobs may be located in the R&D department at a company location, or directly in food production facilities.

Quality Assurance

Digital scale[6]

A common job for those with a bachelor's degree in food science is in quality assurance or quality control. In this work, scientists:

[iv] Salary range reference: Kuhn, Mary Ellen, and Margaret Malochleb. "2017 IFT Employment & Salary Survey Report." The Institute of Food Technologists. Food Technology, p. 30, March 2018.

1. Help choose and design tests that, when passed successfully, mean that the food product is within acceptable parameters. An example is pouched oatmeal with added sugar and flavorings. Salt is often added to the flavor blend. Using reliable objective test measurements, salt levels can be used to determine if the flavor blend was made correctly and added to the individual pouches correctly.
2. Train company employees how to conduct quality assurance tests.
3. Oversee the testing of products.
4. Oversee the generation and completion of required records. These include quality taste tests and appearance results, for example, but mostly consist of governmental regulatory compliance documents. Government regulations pertaining to food have become extensive. These regulations cover ingredient tracking (you must be able to determine which farmer's field a box of cereal came from, for example), equipment sanitation records, employee training records, the amount and type of bacteria that might be found in a production facility, and plans associated with addressing potential hazards, to name a few. Quality assurance personnel, including scientists, spend a lot of time making sure the company is in compliance.
5. Tasting the finished product on a regular basis. This task includes training employees on this process, including how large of a sample to taste, how to taste it, etc. These variables are all very important! Quality assurance technicians may organize finished product group tastings to review production from the previous day. This helps to *evaluate* how the product varies from day to day.
6. Quality assurance personnel may work with other employees to troubleshoot and help solve production problems that result in poor quality food.
7. Make decisions on acceptability of food. Sometimes this involves making a financially unpopular decision to destroy food that's outside the acceptable limits.
8. Management. Usually in a food production facility, a quality manager oversees the quality technicians.

Scientific Evaluation: Using the word *evaluate* to examine something might seem a little strange, but is very common in science, and especially in food science. An evaluation can involve physically looking at, tasting, and smelling a sample. It can also mean that the moisture content, size, and color should be measured using instrumentation. In this sense, *evaluate* can mean many things and is defined by what needs to be examined to achieve the desired result.

Food scientists working in quality assurance can expect an average salary of $57,000 per year.[v] Most people who work in quality will be located directly in production facilities.

Food Laws and Regulations

Scale of justice[8]

You may have heard of the Food and Drug Administration, more commonly referred to as the FDA. This United States federal agency aids in generating regulations that go along with laws passed by Congress, and enforcing those regulations on food and drug products. The USDA, or United States Department of Agriculture, does the same for meat and dairy products, among other things. The USDA and the FDA have some overlapping responsibilities regarding enforcement. The Food Safety Modernization Act of 2011 greatly increased the regulatory burden on food manufacturers.

The generation, passage, regulation, oversight, and enforcement of food laws and regulations require the expertise of

[v] Salary range reference: Kuhn, Mary Ellen, and Margaret Malochleb. "2017 IFT Employment & Salary Survey Report." The Institute of Food Technologists. Food Technology, p. 30, March 2018.

professionals in the industry. They may be employed directly by the government agencies, or by other groups or companies that advise these organizations.

When enforcing regulations, plant visits may be needed, requiring some travel and being in the plant as a not-necessarily-welcome visitor. These visits are usually unannounced. Other scientists may be employed in testing products at the main locations, which will involve extensive analytical equipment training.

Food scientists working in food law and regulation, in particular in government and with a master's degree, can expect annual salaries averaging $87,000.[vi] Required education can vary widely depending on the job.

Research and Development (R&D)

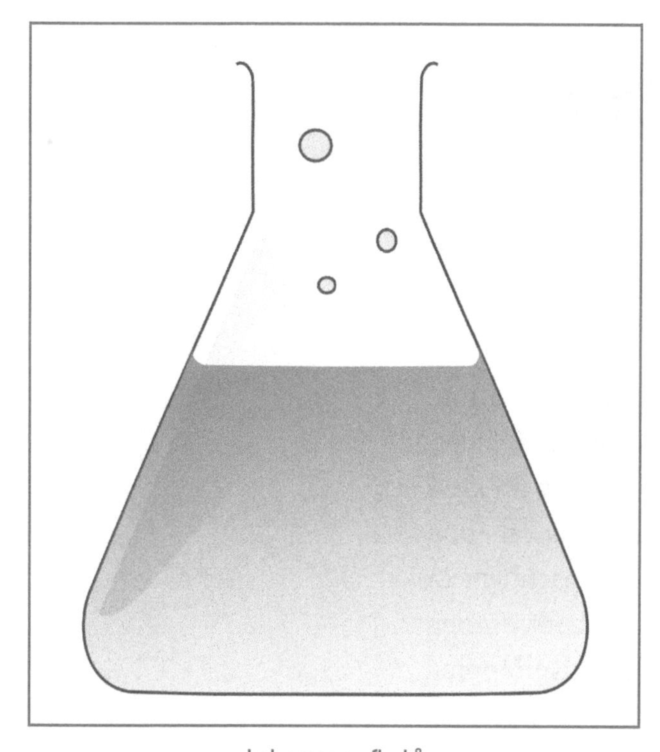

Laboratory flask[8]

An exciting career for food scientists is in developing new food products. Either by yourself or as part of a team, you can take a new food concept all the way from an idea to seeing it for sale in your local grocery store, which can be very reward-

[vi] Salary range reference: Kuhn, Mary Ellen, and Margaret Malochleb. "2017 IFT Employment & Salary Survey Report." The Institute of Food Technologists. Food Technology, p. 30, March 2018.

ing. Usually for this type of work you'd be located at a company headquarters. This means that as testing progresses and the product moves into the start-up of production, frequent trips to the production facility may be required.

As a food scientist working in R&D, you may be working with engineers, sensory scientists, and quality assurance employees. This type of work can be quite challenging, particularly when the product is a brand-new idea that has not been tried before. As a whole, development of new products and all the required steps is referred to as *new product development.*

Large companies employ highly trained and educated scientists who are on the cutting edge of research in a narrowly defined field. These experienced people also can serve to help with ideas when new product development teams run into roadblocks.

Food scientists employed in R&D generally have master's degrees, and sometimes PhDs, but not always. Salaries can range quite widely, averaging $83,000 annually.[vii]

THE AUTHOR'S CAREER IN THE FOOD INDUSTRY

Now that we've reviewed some jobs that are available, let's use my career as an example. My experience has largely been in R&D, employed by larger food companies. A master's degree, which I have, is often required to open the door to these types of jobs, but not all the time. Smaller companies may have different criteria.

General Mills[9]

Gorton's Seafood®, Gloucester, Massachusetts: Product Development and Process Support

My career began working in product development for Gorton's Seafood, which at the time was a division of General Mills—the Cheerios© guys. I worked on developing a new system for the crunchy crumbs that are on the outside of a fish stick (a batter and breader system), which was a very large project.

Gorton's[10]

[vii] Salary range reference: Kuhn, Mary Ellen, and Margaret Malochleb. "2017 IFT Employment & Salary Survey Report." The Institute of Food Technologists. Food Technology, p. 30, March 2018.

proving, and optimizing established products were again part of my daily responsibilities.

Recruiter: Sometimes referred to as a *headhunter*, these individuals are hired by companies to fill job openings. They work from resumes or by actively soliciting qualified individuals through aggressive networking. This process includes reviewing candidate's job experiences and negotiating salaries, and the recruiter typically receives a fee from the hiring employers.[viii]

I also worked on new product development, including with Oreo O's® cereal, Honeycomb®, and the fruit-flavored frosted flakes that ended up in the peach and strawberry varieties of Honey Bunches of Oats®, among other things.

Eventually Kraft bought Nabisco, which owned the Cream of Wheat® and Cream of Rice® brands. I was given primary R&D responsibility to support these brands, which included new product development and troubleshooting established products. I was involved in moving production from the Minneapolis, Minnesota, facility that had been making the products for 80+ years to other locations. This project was one of the most challenging I encountered throughout my career!

The process for new product development is generally as follows:

Marketing refers to a function that deals with advertising the products, as well as determining what consumers like.

1. You (the food scientist) are given a consumer winning *concept* from *marketing*, which is a description of a food product, sometimes with a finished product drawing. A *price point* is usually also available.

2. You work to develop a food that fulfills the concept. This involves considerable work in developing *formulations* (recipes) and laboratory testing, followed by small-scale production, and finally in a full-scale production facility.

3. Consumer tests are conducted at appropriate intervals. A *fulfillment test* is usually the final test before going to market, which involves consumers in the right market. They take the

viii "Recruiter." BusinessDictionary.com. © 2019 WebFinance Inc., All Rights Reserved. http://www.businessdictionary.com/definition/recruiter.html. Accessed 04 February 2019.

food home and use it while comparing it to the concept. They grade the food based on whether or not it was what they had in mind when they first saw or heard the concept.

It's very expensive to launch a new product, especially nationwide. Great care is generally taken to avoid costly mistakes. This will be discussed in more detail later.

4. Starting up the product at the production facility and writing the specifications so that the food produced is always the same and always the best quality.

> ***Price Point:*** A price at which consumers have communicated they'd theoretically purchase a new product. This may be determined during a focus group discussion with consumers from the target market.

Seeing the finished product for sale in the grocery store while you're shopping is a lot of fun. I have a brag box with empty cartons of products I've developed, or was involved in developing.

Brag box[12]

Planters, East Hanover, New Jersey: Product Cost and Process Optimization

After more than six years in Battle Creek, I transferred within Kraft to the main Nabisco research facility in East Hanover, New Jersey. It was the first time I was located at a primary facility—with hundreds of professional food scientists, including some world-renowned people who were at the absolute top of their areas of expertise. It was a new experience having them around. They could answer just about any question you had, and as a bonus, they were only down the hall!

I was also able to purchase Kraft and Nabisco products at a steep discount in a company store, and I even got some food products for free. Researchers frequently moved their leftover test products into the hallways and people could take them home—large amounts of packaged Cheese Nips®, Oreo's®, Triscuits®, etc. It was definitely not a good scenario for maintaining a healthy weight! Luckily, another perk was a good on-site gym.

Planters[13]

> Cost reduction can come from finding less expensive ingredients, as well as making improvements in the production process.

I worked on cost reduction projects, but was also introduced to process automation, which was something that I grew to really enjoy. Data from on-line (on the production line) gauges (meters or sensors), along with process control software, was used to optimally control a cooking process. The on-line sensors continuously monitored the temperature, color, and raw material, and finished product's moisture content. Software and associated sophisticated control algorithms used the data to predict what conditions were needed to produce an optimal product. These types of systems can save considerable money for the manufacturer and help produce a very consistent product for the consumer.

I ended up living in Pennsylvania for over two years, driving 140 miles round-trip each day to get to work. Then a recruiter representing Kellogg's® called with an invitation to move back to Battle Creek, Michigan. Returning to a 10-minute commute sounded very good, and away I went with my family!

The Kellogg Company, Battle Creek, Michigan: Product Development and International Troubleshooting

Kellogg's[14]

Kellogg's owns many brands, including all the Keebler™ varieties of cookies and crackers and Cheez-Its®. At Kellogg's, I worked on these types of products for the first time. My product development role was as I described earlier. At this time

I discovered that as opposed to new product development, I preferred production troubleshooting, applying my knowledge and my experience.

I had the opportunity to troubleshoot more problems when I transitioned to a new job within the international division. I spent considerable time on the phone with Kellogg Company groups in other countries, frequently at odd times of the day, at least for me! I also traveled to Kellogg's production facilities throughout the world and helped to troubleshoot problems. I visited production facilities in Japan, South Korea, South Africa, and Mexico. These trips were interesting and meeting new people was great, but all food production facilities look pretty much the same from the inside, regardless of their location in the world. Language barriers add to the challenge. At times, I didn't speak their language and they didn't speak mine, but we both spoke a third language!

> Your interests may change as you grow and mature. You've probably noticed this normal progression already! The activities you enjoyed when you were younger may not be the same now, and the ones you enjoy now may not be what you enjoy next year. The key is finding a job or career that you enjoy and at which you can make a decent living.

> Rarely do you get the opportunity to bring your family with you on trips. On an overseas trip—where you may have days off to let you recover from jet lag—sightseeing is not so much fun when you're doing it by yourself. When traveling, bring someone with you, if you can. Paying the cost of an extra plane seat is worth having someone you can share your experiences with!

Post Consumer Brands/Malt-O-Meal/MOM Brands, Asheboro, North Carolina: Senior Scientist/Site Scientist

After almost five years at Kellogg's, another recruiter called about a job in a southern state, near where my wife and I had considered eventually retiring.

The position involved troubleshooting and solving problems (termed *problem-solving* in the industry) at a production location—what I enjoyed the most. I used my experience to help fix whatever daily issues arose from day to day, and worked to optimize and improve formulas and the manufacturing process. I was also able to use my process control experience and introduce some new methods to the company.

Malt-O-Meal[15]

Beakers & Bricks, LLC, Asheboro, North Carolina: Owner and Author

Following almost nine years as the Site Scientist for the Asheboro ready-to-eat cereal production facility, I decided to make the jump and work for myself. I have run side businesses and enjoy the challenge and sense of satisfaction in providing a service and products that people value. I didn't have enough time in the day to do both anymore, so I decided to dedicate my time fully to my own business.

Beakers & Bricks, LLC, publishes educational materials, including this workbook, along with offering consulting services.

WHAT DO YOU THINK?

When considering your career and employment opportunities, you'll have a choice of the size of company you want to work for, as well as whether they're privately held or are a public company. My experiences can give you an idea of what it is like to work in different situations. Work environment can make all the difference in whether you enjoy going to work or dread hearing the alarm sound each morning.

Regardless of where you work, you should never "burn your bridges" on your way out the door from a place of employment. Part on as friendly and professional terms as possible every time. While always a good policy, it's particularly so in the food industry. During my relatively short experience, companies have frequently merged, or been purchased and sold. For example, note that the production facility job I had in Asheboro shows three different company names. It started out as Malt-O-Meal, Inc., which changed its name to MOM Brands, Inc., and was later purchased by Post Holdings, Inc. and, together with Post cereals, was formed into a new division of that company called Post Consumer Brands.

You're likely to run into some of the same people as your career develops. For example, my manager while at Malt-O-Meal/Post used to sit across the hall from me when we both worked in research at the Kellogg Company in Battle Creek, Michigan. And when Post Holdings, Inc. purchased MOM Brands, previous colleagues from when I worked at Post the first time became my co-workers again. I hadn't seen them for 13 years, but thankfully, there were hugs and warm feelings all around, for the most part. It could have been very different, so keep in mind that you never know when an old boss may become your new boss!

JOURNALING IDEA

Imagine that you're a professional food scientist having just completed 20 years of working for several food companies. Write a two-page fictional account of your favorite new product development project. Include details such as who you worked with, what types of challenges you had when developing the food, what the package looked like, and how successful it was in the marketplace.

Use complete sentences, and share your story with someone else.

If you are interested, go to www.beakersandbricks.com for further stories, insights, and videos from Dale's experience in the food industry.

CHAPTER REVIEW

1. Look through the cupboards in your pantry or kitchen, as well as the refrigerator and freezer. Inspect the packages to see who's selling the product. Evaluate at least five different products and make a list, including the company names that produce them. You can find the company information on the side panel, and often on the container front.

2. Research two food companies and their websites online (Nestlé, Kellogg's, or General Mills, for example). You can choose any that interest you. If you can't think of any, go to your kitchen cupboard and look at the side panel of something in there. A section will indicate "distributed by" and give a company name. Use the forms on the following pages to record information about the companies, the products they make, and job opportunities that may exist there.

Store Brands: Sometimes called private label brands or generics, it's likely that store brand products were produced by a different company for the store. In these cases, the product is distributed by the store, not manufactured by them. This arrangement is called co-manufacturing—a production company produces and packages products for other companies in addition to their own brands. Some companies only manufacture products for others and don't sell any of their own brands.

Note: For any written activity in this book, please use complete sentences.

Company #1

1. Company name:

   ____

   ____

2. Where are they headquartered?

   ____

   ____

3. Where are their manufacturing locations? If you were employed as a quality assurance technician, these are the locations where you might live.

   ____

   ____

4. Where's their R&D location? If you were employed developing new products, you'd likely live here.

   ____

   ____

5. Look at the different brands they produce. List three.

   ____

   ____

   ____

6. Find the link to their careers page. What technical jobs are available, especially those related to food science?

   ____

   ____

7. What are the job duties for those positions?

   ____

8. What are the requirements for those jobs?

9. What are the salary ranges? These will sometimes be listed directly on the website.

Company #2

1. Company name:

2. Where are they headquartered?

3. Where are their manufacturing locations? If you were employed as a quality assurance technician, these are the locations where you might live.

4. Where's their R&D location? If you were employed developing new products, you'd likely live here.

5. Look at the different brands they produce. List three.

6. Find the link to their careers page. What technical jobs are available, especially those related to food science?

7. What are the job duties for those positions?

8. What are the requirements for those jobs?

9. What are the salary ranges? These will sometimes be listed directly on the website.

Fun People: One development team at Post developed the Great Grains® line of breakfast cereals that consist of flakes of different types, dried fruits, granola, and sometimes walnuts, pecans, and almonds. They made a T-shirt to celebrate their success that said "I work with a bunch of flakes, fruits, and nuts." Pretty funny!

CHAPTER 2
A SCIENCE PRIMER

While this workbook series isn't intended to be a chemistry class and it doesn't require extensive use of scientific principles, a basic knowledge of the relevant terminology is needed in order to properly discuss the concepts involved. The following information should be sufficient for a novice to successfully work through the Edible Knowledge® workbooks.

ELEMENTS AND THE PERIODIC TABLE

Everything in nature is made up of elements, which are the basic units that current science can describe as being unique from everything else. Elements are also referred to as *atoms* (for example, 1 atom of hydrogen). Elements such as hydrogen, oxygen, iron, and helium are all made up of protons, neutrons, and electrons. Protons and neutrons together can be thought of as taking up the sun's location in our solar system, with electrons being the orbiting planets.

The number of these components and the way in which they naturally arrange themselves is what makes them more or less *reactive* with other elements, and it's also what makes up their chemical characteristics.

Chemical reactions happen when different elements are mixed together and chemically combine to form different materials. *Reactants* are what combine together, and *products* are the result—usually more than one. In equation form, reactants are always on the left and products on the right:

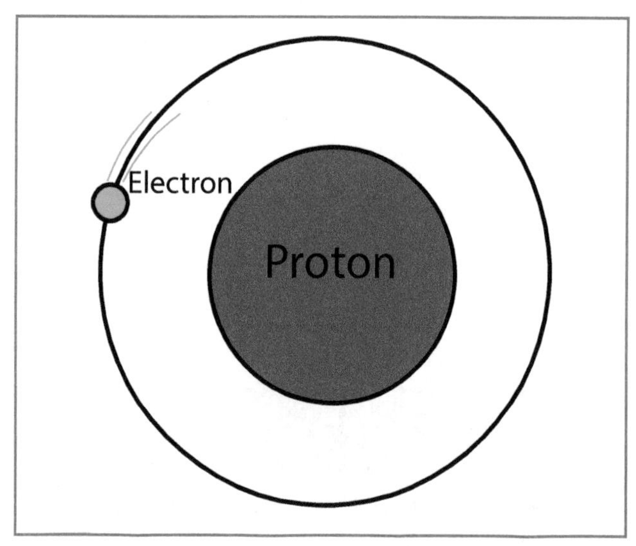

Hydrogen atom[16]

Reactant A + Reactant B = Product(s)

Reactions can be slow or fast. Sometimes they generate heat, while other times they absorb heat. The interactions are primarily associated with the electrons and the way that they're moving around in their *orbitals*.

Orbitals: Another name for the paths and areas occupied by electrons in an atom. The world of chemistry is complex and fascinating, and that makes food science challenging!

However, electrons move very, very fast. Deeper levels of structure associated with the protons, neutrons, and electrons themselves are known, but they're beyond the scope of this workbook.

The Periodic Table was first organized in the late 1800s by Dimitri Mendeleev, and has been refined many times since then. The table's primary purpose is to organize elements in ways that correspond to how they react with one another, which helps predict reactions. Elements have a one- to three-letter symbol by which they're referred to, both on the Periodic Table and in chemical equations. The table lists the elements

A college chemistry professor was asked by a student why a particular chemical reaction happened the way it does. His response was, "We don't know. That's a religious question!" He explained that observations from nature show us how chemicals react, and we can even make rules that describe and categorize these reactions to where we can predict what will happen, but we don't know "Why?" He said that perhaps the basic elements love each other, or hate each other—we simply don't know. When you think about it, he was right!

in rows (periods) and columns (families). Separate from this, the elements are listed by the number of protons they contain. Scientists occasionally discover new elements, but as of 2018, the Periodic Table lists 118 elements from hydrogen (1 proton) to oganesson (118 protons).

The table's periods and families have other sections or groups that are composed of elements that react in a similar fashion. Some of these groups include the transition metals, alkali metals, alkali-earth metals, and noble gases, among others. They're grouped this way in response to many scientific observations that have resulted in rules describing how elements may react with one another. These groupings help to predict the type and strength of chemical reactions. Below is one depiction of the Periodic Table of the Elements.

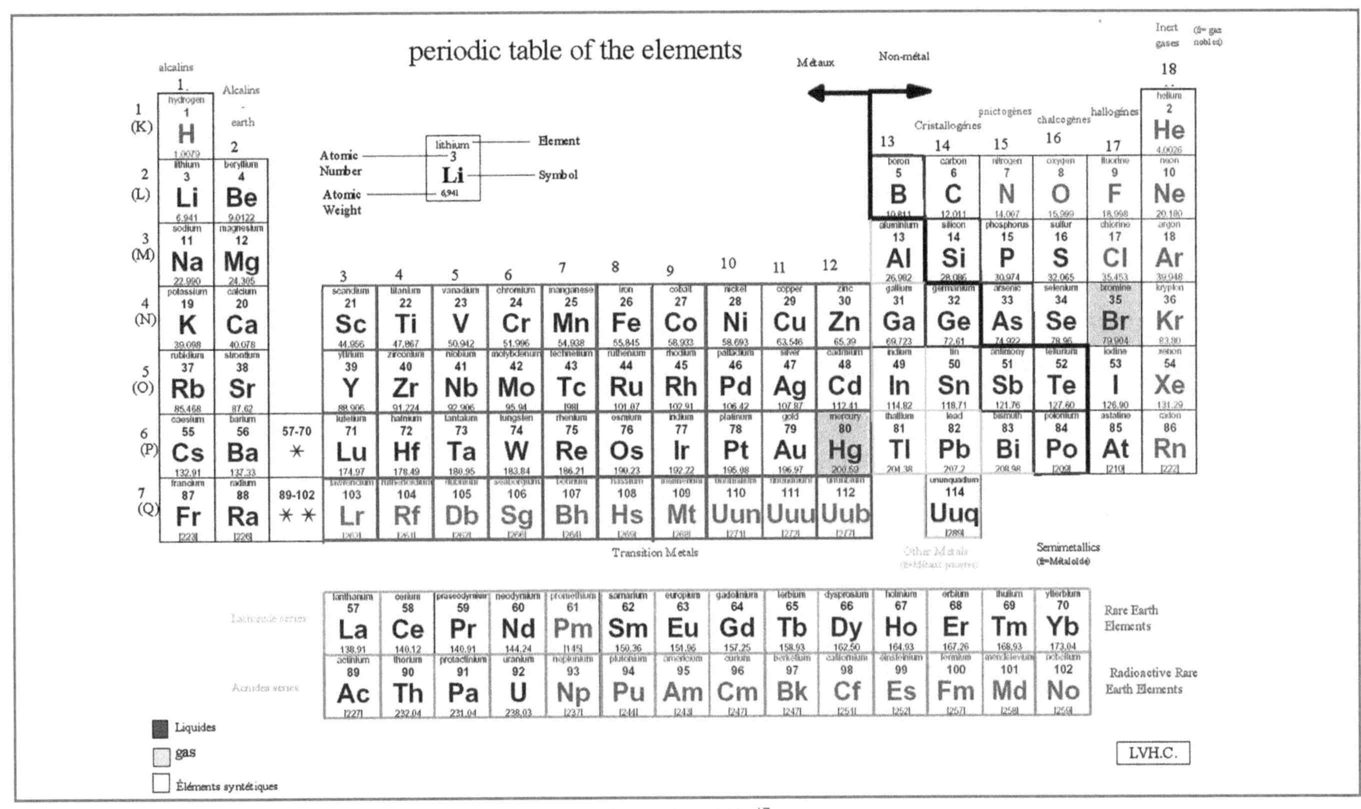

Periodic Table[17]

CHARGES AND MASS

A magnet has one end that's positively charged and another end that's negatively charged. Have you ever tried to force two ends of a magnet with the same charge together? They repel one another. However, if you place magnets with the opposite charges near each other, they attract one another. Depending on the charge strength, this magnetic attraction can be strong or weak.

Protons are positively charged, neutrons are neutral (they essentially have no charge), and electrons are negatively charged.

Protons and neutrons both have substantial mass, or substance, while electrons have very little mass. Mass is different from weight in that the mass of an object doesn't change, but weight will change depending on the gravitational force. For example, an astronaut practicing on the Earth weighs more on Earth than he will on the moon because the gravitational force on the moon is significantly less than that on Earth. The astronaut's mass, however, doesn't change. Mass vs. weight is an important concept, and one of the more difficult to grasp!

CHEMICAL BONDING AND BOND STRENGTH

When elements interact with one another, they may form a new substance called a *compound*. A compound can consist of any number of elements chemically bonded to one another. However, the smallest set of them that can't be split without making something else is called a *molecule*. For example, one molecule of water is made up of 1 oxygen atom and 2 hydrogen atoms. If you tried to make it any smaller, you wouldn't have water—you'd have hydrogen and oxygen.

When elements combine to form a molecule, it happens through different types of chemical bonds. Without bonding, only individual elements would exist, which would really be strange. Chemical bonds differ in strength by type. Put simply, the bond's strength depends on the outermost electron's nature and availability.

Ionic bonds exist when an electron is transferred between elements. Covalent bonds exist when an electron is shared between elements. Ionic and covalent bonds are the strongest bonds, but not all science professionals agree which is the strongest. In practice, the strength of any bond is very complex and depends on the total compound and the environment in which it exists.

Other types of bonds include hydrogen, dipole-dipole, and Van der Waals. All of them are the result of elemental electrical charge. The electrons themselves aren't transferred or shared, which makes these types of bonds significantly weaker than ionic or covalent bonds, but they are still very important in describing, understanding, and predicting how substances will react.

Hydrogen Bonds

Hydrogen bonds are particularly important in the food industry. Hydrogen bonding is characteristic of water (H_2O) interactions with other molecules of water, and also with molecules of other compounds.

Our understanding of hydrogen bonding helps explain why water is crucial to life on Earth, and how it plays such an important role in our food.

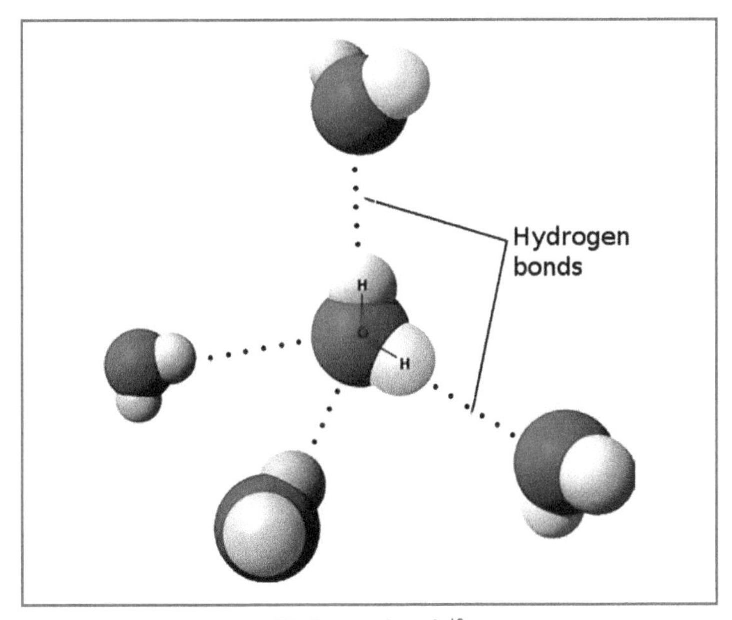

Hydrogen bonds[18]

A water molecule has 2 hydrogen atoms and 1 oxygen atom. The molecule's shape ends up looking lopsided, as you can see. The red sphere is the oxygen atom and the 2 white ones are the hydrogen atoms. The molecule itself has a negative charge on the oxygen side and a positive charge around each hydrogen atom.

The result is a bipolar molecule that's similar to a magnet. When many water molecules are present, the hydrogen will form bonds with one another. These hydrogen bonds exist continuously throughout any body of water. There is additional discussion on hydrogen bonding in the Edible Knowledge® workbook titled *Introduction to Food Science: Water*.

Bipolar: An object with a positive and negative side—that is, two poles.

Relative Bond Strength

Although scientists disagree to some extent, today's understanding of bond strength is simplistically depicted in the figure titled "Relative Bond Strengths."

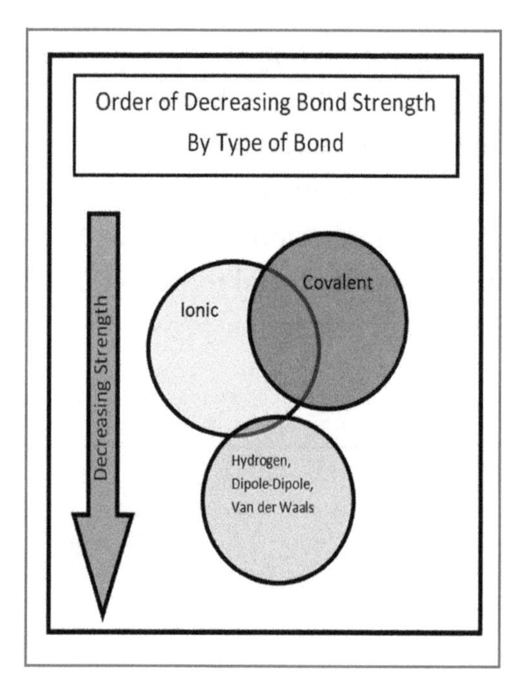

Relative bond strengths[19]

MOLECULAR WEIGHT

The individual atoms of elements combine to form compounds of various types and sizes. The molecular weight of a *mole of any substance* is the sum of the masses of the number and type of atoms that make up the molecule, multiplied by Avogadro's constant, which is 6.022×10^{23}.

A *mole* is a unit of measure of countable things. Often it's compared to a dozen, as that word describes a specific number of countable items—twelve. In fact, a mole is sometimes referred to as a *chemist's dozen*, and instead of twelve, it contains 6.022×10^{23} of whatever we're looking at. For example, one mole of hydrogen is 602,200,000,000,000,000,000,000 atoms of hydrogen. That's a bunch of atoms! Using mole terminology allows chemists to discuss quantities of atoms or molecules necessary for conducting experiments without having to use very large, cumbersome numbers.

Molecular weights can vary widely. For example, water is composed of 2 atoms of hydrogen—the smallest atom—and 1 atom of oxygen. By comparison, 1 average molecule of a natural starch can have thousands of glucose molecules formed into a chain. Each glucose molecule is also much larger than a molecule of water.

MEASUREMENTS

With the United States being the exception, almost all of the rest of the world uses the metric system for expressing physical measurements, including temperature, mass, and length. As a result, the scientific community throughout the world, including the U.S., uses the metric system...but not always!

It's important for you to know that food science involves both science and engineering. Since engineers in the United States often use its standard system instead of the metric system, food scientists need to be able to convert from one to the other and be able to comfortably use both systems.

Temperature

Temperature is expressed in degrees of Celsius rather than Fahrenheit. The conversion factors are below:

Fahrenheit = Celsius × 1.8 + 32
Celsius = (Fahrenheit – 32) ÷ 1.8

The symbol ° following a number is read *degrees*, and the scale is indicated as either an "F" for Fahrenheit or "C" for Celsius. For example, 45°F is read "forty-five degrees Fahrenheit," or "forty-five degrees F." Another example, 24°C, is read "twenty-four degrees Celsius" or "twenty-four degrees C."

While altitude, pressure, and impurities affect freezing and boiling points, some common data points are presented below to help you get used to the differences between the Fahrenheit and Celsius temperature scales.

	Freezing Point of Water	Boiling Point of Water	Cold Day	Nice Day	Hot Day
Degrees Fahrenheit	32°	212°	10°	75°	95°
Degrees Celsius	0°	100°	-12°	24°	35°

Fahrenheit vs. Celsius[20]

My youngest son plays online games with people from around the world. I heard him and one of his online friends from a European country talking about temperatures once and realized the discussion was about different scales. 32°F and 32°C are two very different scenarios when experienced in the real world. The first is ice and snow, and the second is a nice day at the beach!

Fahrenheit sign[21]

Celsius sign[22]

Mass

The metric system uses milligrams, grams, and kilograms, while the system commonly used in the United States uses ounces and pounds. The conversion factor from pounds to kilograms is as follows:

1 pound (lb.) = 0.45359 kilograms (kg)

In food science, most often weight or mass measurement uses the metric system.

1 kilogram (kg) = 1,000 grams (g) = 1,000,000 milligrams (mg)

Formulas (recipes) developed by food scientists for use in a manufacturing plant are frequently in grams and kilograms, which need to be converted to pounds and ounces for actual use in the plant, although many times both are used. It's important to make these conversions correctly, or you could end up with some really funky food!

Length

The metric system uses millimeters, centimeters, and kilometers, while the standard (U.S.) system uses inches, feet, yards, and miles. The conversion factor for centimeters to inches is below:

1 inch (in) = 2.54 centimeters (cm)

Volume

The metric system commonly uses just milliliters and liters, while the U.S. system measures volume in fluid ounces, pints, quarts, and gallons. The conversion factor for liters to gallons is below:

1 gallon (gal) = 3.79 liters (L)

Solutions: Molarity vs. Molality

Chemical concentrations of solutions of a *solvent* and a solute are commonly measured by molarity and molality:

Molal concentration = moles of solute ÷ kilogram of solvent
Molar concentration = moles of solute ÷ liter of solution

> ***Solvents, Solutes, and Solutions:*** Solutes are dissolved in a solvent, which together make a solution. For example, in a solution of sugar dissolved in water, sugar is the solute and water is the solvent.

Both molality and molarity are used extensively in different situations. Understanding the difference is required to understand principles that will be presented throughout this workbook series.

pH

An objective indicator of the acidity or basicity of a solution is pH. This measurement is used throughout food science. A neutral (in the middle) pH is 7.0—pure water has a pH of 7.0. On the pH scale, a lower number indicates acidity and a higher number means the solution is basic. *Acidic* means more hydrogen (H^+) ions are in the solution, while *basic* means more hydroxyl (OH^-) ions.

pH	Examples of solutions
0	Battery acid, strong hydrofluoric acid
1	Hydrochloric acid secreted by stomach lining
2	Lemon juice, gastric acid, vinegar
3	Grapefruit juice, orange juice, soda
4	Tomato juice, acid rain
5	Soft drinking water, black coffee
6	Urine, saliva
7	"Pure" water
8	Sea water
9	Baking soda
10	Great Salt Lake, milk of magnesia
11	Ammonia solution
12	Soapy water
13	Bleach, oven cleaner
14	Liquid drain cleaner

pH scale[23]

WHAT DO YOU THINK?

Who is the most famous scientist you can remember? What was that discovery or research that made that person famous? Do you think there are many more things to discover?

JOURNALING IDEA

Why do atoms react together the way that they do? Describe—in as much detail as you can—the difference between "how" and "why" in scientific knowledge and progress.

Sometimes people, even scientists, use the phrase *settled science* to refer to a body of knowledge that's believed by them to be proven and complete. Do you agree with this? Why or why not? Use complete sentences.

CHAPTER REVIEW

Write short essay responses to each of the following situations:

Imagine you're working in a laboratory where standard and metric measurements are both used. Describe what precautions you'd need to take to avoid making a mistake.

Search on the internet and read about the NASA loss of a Mars orbiter that occurred in September 1999 (http://www.cnn.com/TECH/space/9909/30/mars.metric.02/). Describe what it must have been like for those on the project to realize the mistake after years of work and tens of millions of dollars spent to bring the project to fruition.

CHAPTER 3
FOOD PROCESSING

***Processed food.* Depending on who** you're talking to, this phrase could almost be considered swearing. Processed food is viewed negatively by many people, and frequently is treated as the reason for the increase in obesity and other diseases. In the basic sense, however, almost everything we eat has been "processed" in some manner. Raw produce, such as a carrot, is harvested. The green tops are cut off and the rest is washed. The carrot is now a processed food. At home, you might take carrots from the bag, peel, cut, and steam them before eating them. That's additional processing.

Hostess Twinkies®[24]

Of course, when most people think of processed food, something like a Hostess Twinkie® or a frozen chicken pot pie comes to mind, not a peeled carrot. There's a progression (or continuum) of processed food, from a minimal process—such as preparing fresh vegetables—to something highly processed.

Sliced carrots[25]

Almost all the foods we eat are processed, so we shouldn't be afraid of eating a processed food. In fact, food science and food manufacturers continue to work together to ensure food is safe, as well as reduce waste through advancements in food processing. The key is to eat many types of food, but in moderation! For more on this, refer to the section called *My Philosophy*.

In general, the more highly processed the food, the more readily the body can turn it into glucose, since it's already been partially digested. I know that sounds disgusting, but that's the way I think of it! Consider what your body does to food when you eat. You bite and chew the food, breaking it up into small pieces and mixing in *enzymes* from your saliva that begin to break down the food.

The food is transferred to the stomach, where the breakdown process, or digestion, continues chemically (with acids and enzymes) and mechanically (stomach wall muscles churning the food). The goal of digestion is to break food down into molecules the body can use, such as the individual amino acids that come from protein, and glucose, which comes from carbohydrates. For example, the starch from a piece of popcorn is largely broken down into individual glucose molecules in our stomach.

Enzymes are complex proteins that facilitate, or promote, biochemical reactions. Another term commonly used is that enzymes *catalyze* these reactions.

Popcorn[26]

Depending on the production process, the mechanical, artificial "digestion," such as grinding or mixing, involved in producing a food can have many different degrees of extent. If there's a lot of mechanical work done to make the food, it's likely easier for the body to get the glucose into the bloodstream, thereby also making it easier for the body to turn it into fat.

For example, compare shredded wheat cereal with cake flour. Both are made from wheat, but the wheat is processed very differently to be able to make those two products. To make cake flour, wheat is ground into a fine flour and the bran and germ are removed. The wheat in shredded wheat is essentially boiled whole, smashed between two rollers, and it's done. In this example, the wheat is more highly processed to make cake flour than it is to make shredded wheat. To be clear, cake flour is a more highly processed food than shredded wheat cereal.

Raw food is processed so it's easier to use. In fact, many raw foods, such as wheat, can't be consumed unless processed in some way!

Generally, the more processed the food is, the easier and more convenient it is to use and consume. But because its form is closer to what the body can use, the food is more easily converted by the body into fat. This includes finished products that can be consumed directly from the package—a relatively new development in food history. This is convenient, and in fact is the origin of the term *convenience foods!*

Once processed, these foods need to remain safe to eat—and taste good—until they can be purchased and consumed. Considerable application of food science is involved to make this happen. Ingredients are added that help to preserve the flavor, texture, and color, or to prevent the growth of something that would ruin the food, such as mold. These ingredients collectively are called *preservatives*.

MY PHILOSOPHY

Most of my career has been involved in producing foods that would be considered moderately to highly processed. Some might find it interesting that my personal diet doesn't include many of the foods that I've worked on. Remember, since convenience foods tend to be more highly processed, they are more easily converted into glucose, and therefore fat. Many nutritionists and doctors believe they should be consumed only when convenience is needed, or perhaps when you want an occasional treat. For example, on a long road trip or when camping, it's convenient to eat crackers or cookies out of a box, or to eat cereal from a box.

For myself, I believe that it's best to eat food as close to the natural state as possible. As time goes on, scientists are also discovering that eating food as close to its natural state as possible is best for the human body. But don't get me wrong—I love a good box of Cheez-Its® just as much as the next guy! And, in fact, I even worked on them as part of my career. I just try to eat them as a treat, not as my regular food. It's always a good idea to learn about what you're eating and take responsibility for what you put in your body.

As an aside, this philosophy is one reason why I'm not in favor of having the government regulate food availability. Forcibly removing a desired product from the marketplace because in excess it may be bad for people seems to be the wrong way to go. I see it as an attempt to force people to make good decisions. If you want to eat a box of Ding Dongs®, go for it! Just don't blame anyone else for the health consequences you may experience.

AN EXAMPLE OF A COMMERCIALLY PROCESSED FOOD

Let's look at an example of a processed food and what happens to the raw materials as it's produced.

Frosted Shredded Wheat Cereal

Vintage Shredded Wheat advertisement[27]

The ingredient line from the side panel of a bag of Malt-O-Meal Frosted Mini-Spooners® is shown at right.

Let's break down each ingredient in detail to determine how it's processed and how it ends up in a bag of shredded wheat cereal.

Ingredients: Whole Grain Wheat, Sugar, Contains 2% or less of: Gelatin, BHT (to preserve freshness).

Vitamins & Minerals: Vitamin B1 (thiamin mononitrate), **Vitamin B2** (riboflavin), **Niacin** (niacinamide), **Vitamin B6** (pyridoxine hydrochloride), **Folic Acid, Vitamin B12, Reduced Iron, Zinc** (zinc oxide).

Contains Wheat Ingredients.

Shredded wheat ingredients[28]

Green wheat in the field[29]

Wheat

1. Wheat is a grain that's grown in a field and looks essentially like grass, with large bunches of wheat *berries* at the end of the stalk—the individual kernels of wheat.

Wheat, ready to harvest[30]

2. Once the wheat has dried to the proper moisture content, it's ready to harvest. This can be tricky, since most of the drying should take place in the field. If the weather doesn't cooperate, entire season's crops can be ruined.

Wheat harvest in Germany[31]

3. The wheat is harvested with large *combines*, or machines that combine several actions. These can be modified depending on the crop being harvested, but for wheat they typically strip the stalk and husk from the wheat berry, store the berry in an onboard hopper, and finally grind and spread the stalk and husk refuse on the field as fertilizer. Trailers pull up alongside the combine (even while it's still harvesting) and the combine transfers its load of wheat to the trailer.

Watch this cool video showing wheat being harvested...and some music to go along with it! https://www.youtube.com/watch?v=zX8K1OpBCj4

4. The wheat may be stored for some time in silos, but eventually it's transported to a mill, where it is further processed to remove any sticks, stones, other types of grain (such as oats or barley) that may be mixed in, as well as any other foreign material. In the case of ready-to-eat shredded wheat cereal, the wheat is typically purchased by cereal manufacturers at this stage of processing.

Smaller-scale wheat cleaner. https://www.youtube.com/watch?v=FFfZ3dzUXHs

Sugar

Chemists use the term *sugar* to refer to simple carbohydrate molecules like glucose, fructose, sucrose, lactose, and maltose. Table sugar, or the granulated white stuff we all love so much, consists almost entirely of sucrose. Sucrose is a *compound* sugar composed of a one-to-one combination of glucose and fructose, chemically joined together. Compound sugars, also known as *disaccharides* or *polysaccharides*, are combinations of more than one basic sugar. More on this in Chapter 6.

As the main ingredient in the frosting on frosted shredded wheat biscuits, sucrose can be derived from many different sources, including sugar cane and beets. For the purpose of this discussion, we'll evaluate how granulated sugar is prepared from sugar beets.

It may seem strange to get sucrose out of sugar beets. Sugar beets are a root vegetable, like the more familiar red beets. However, they become lighter in color, similar to a parsnip. It's definitely ironic that sugar—what brings children immeasurable pleasure—can be derived from something they stereotypically avoid: vegetables! Nevertheless, they can produce a lot of sugar.

The following is the basic process of extracting and refining the naturally present sucrose from sugar beets:[ix]

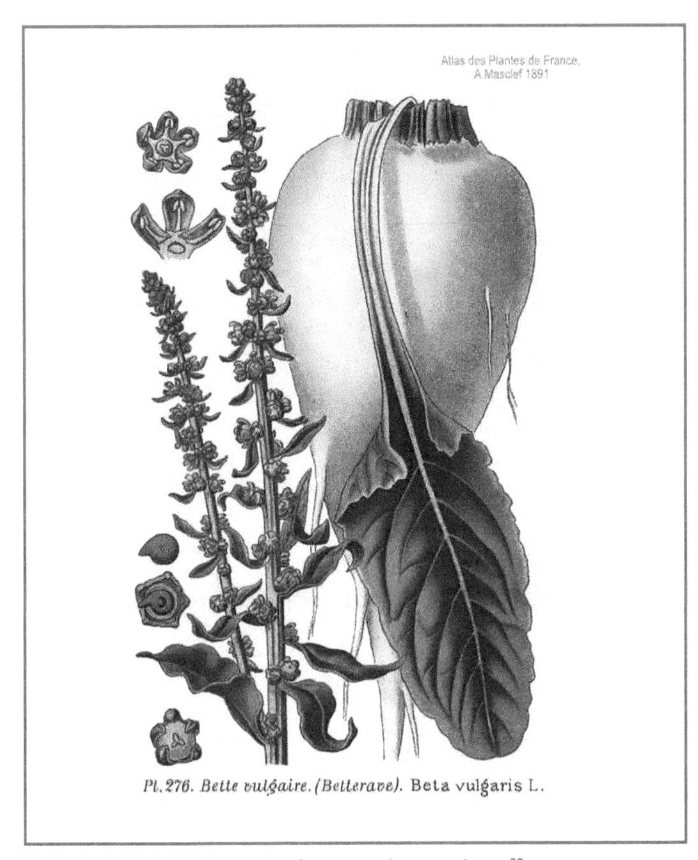

Drawing of a typical sugar beet[32]

1. Sugar beets are grown and harvested by large machines that loosen the soil and separate the soil from the root.
2. The beets are washed and cleaned so that all the leaves are off, and no dirt or other foreign material remains. Beets are floated in water to separate them from any stones that came along for the ride when the beets were harvested.
3. Once cleaned and topped (the tops, or leaves, removed), the beets are sliced into very thin strips or disks to increase the available surface area. These slices are sometimes called *cossettes*.
4. The slices are soaked in a diffuser while moving slowly through hot water in a *counter-current* fashion. In other words, the beet slices/strips that have been in the cooker

ix "Growing & Processing Sugarbeets." Michigan Sugar Company, © 2015 Michigan Sugar. http://www.michigansugar.com/growing-production/from-seed-to-shelf/. Accessed 02 February 2019

the longest move through the freshest water. This allows for the maximum diffusion of beet juices into the water by moving from the area of highest to lowest concentration. The counter-current methodology is used frequently in the food processing industry to efficiently extract, heat, or cool material, among other things.

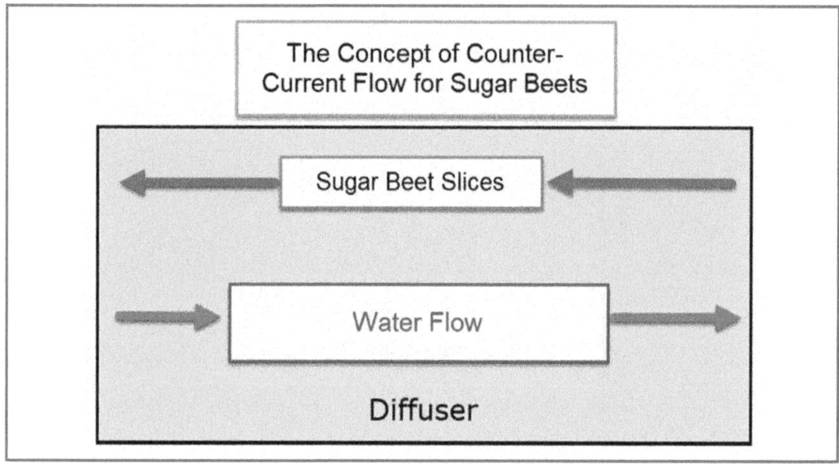

Counter current flow[33]

5. The beet pulp that's left after drawing off the juice may be pressed to further remove additional juice. What's left after that may be used for animal feed, fertilizer, or other products.
6. The raw juice that results from this process has color and impurities in it, including sugars other than sucrose. Various methods are used to remove the impurities and color. In addition to filtration, *lime* and carbon dioxide (CO_2) are used in the purification process. As lime and carbon dioxide are percolated through the juice, a *precipitate* forms from protein and other impurities. This precipitate either settles out or is *centrifuged* out, resulting in a clear juice!

Precipitate refers to a solid that's generated through a chemical reaction.

Centrifugation: A centrifuge is a device that spins at a high rate of speed, magnifying the effect of gravity. Using these devices is useful for separating out suspended solids in a liquid.

7. The resulting juice is mostly sucrose. It may be purified further, or it can be concentrated and crystallized

at this point. This involves removal of water by several methods. The most common method is by boiling the liquid under a vacuum. The vacuum allows for boiling and evaporation at lower temperatures, requiring less energy input and reduced thermal abuse of the sucrose.

8. The resulting syrup/crystal mixture is centrifuged to remove the crystals.
9. These crystals are dried, separated by granule size (Have you heard of "powdered" sugar?), and packaged in small bags to be sold to the consumer, in larger bags for restaurants, and even in rail cars to be used by food manufacturers.

Years of experience and associated expertise are involved in each step. The companies that produce sugar utilize years of experience to get it right and to produce sugar at an affordable price for the consumer while still making a profit for themselves.

A short video showing sugar produced from sugar beets.
https://www.youtube.com/watch?v=VRZX1bAnbes

Another video showing production from sugar cane:
https://www.youtube.com/watch?v=jCKt02NGjfM

Gelatin

Gelatin production can be a little scary (and perhaps even a bit gross!). Some of you may not want to eat gelatin-based products for a while after reading this and learning how gelatin is made. You shouldn't be worried, though, because gelatin is entirely safe. You just may not want to think about where it comes from…

Gelatin itself is a protein that's derived from collagen. Collagen is found in natural connective tissues such as tendons, ligaments, bones, and skin. The primary raw materials used for production of gelatin are cattle bones, cattle hides, and porkskins, and to a lesser extent, poultry and fish.[x]

The following is the general process for producing gelatin. We're using pork skins in our example.[xi]

1. The pork skins arrive at the gelatin production facility where they're inspected for quality and chopped into small pieces.
2. Depending on the animal part's source, they may be

[x] "GMIA Gelatin Manual." Gelatin Manufacturers Institute Of America, 2012. http://www.gelatin-gmia.com/images/GMIA_Gelatin_Manual_2012.pdf. Accessed 02 September 2017.
[xi] Ibid.

treated with lime (a basic, or caustic, high pH liquid or powder) or an acid solution. Some processes can take up to several weeks. The tasks can be done in batches and may be done in several rounds in order to extract as much gelatin as possible. In all cases, the pH, time, and temperature are carefully controlled to achieve the quality of gelatin desired.

3. The various "extractions" are kept separate. These extractions can be mixed and matched to achieve a certain gel strength, as well as other qualities that may be desired by a particular customer.
4. After extraction, the resulting dilute gelatin solutions are filtered, deionized, and concentrated. Concentration can be accomplished by evaporation under vacuum conditions.
5. The resulting dispersion is chilled, resulting in a *gel* (think concentrated Jell-O®).
6. These gels can be cut and dried, which results in the finished gelatin. The product is almost entirely flavorless and odorless, but will have a slight yellow or amber color.

Gels are important throughout the food industry and usually consist of protein, and sometimes sugar. This gel is extruded, or forced through small holes, creating long ribbons similar to spaghetti.

The traditional method for extracting gelatin, when it's not done in a factory or plant, is to skip the acid and base step and boil the animal product for an extended period of time. This is how individuals created gelatin in the past, and in medieval times, it was used to encase food so it wouldn't spoil quickly.

An excellent video showing gelatin manufacturing by one manufacturer, including extrusion. It also shows another type of gelatin, called *leaf gelatin*. **Note that gelatin is spelled** *gelatine* **in some English-speaking countries other than the United States.**
https://www.youtube.com/watch?v=uf0uEWGWLgg

Vitamins

We'll skip the description of how vitamins are manufactured. Each has a production process, but it isn't necessary to provide the details. The different types of vitamins are added together and incorporated into a dry powder blend, which is then added to the cereal, as will be described in the next section. Vitamins are added to products primarily as a marketing tool. That is, the product can be advertised as having "9 essential vitamins and minerals," for example. In some cases, the vitamin fortification might be integral to the concept, such as in Kellogg's Product 19®, which has 19 vitamins and minerals. Other times, vitamins are required to be added by government regulation. When you can have great tasting food that doubles as a vitamin pill, it's a very good thing!

> One note about making cereal is that you have to make a LOT of it in a short time for it to be a viable business venture. Rail car quantities are usually required due to this constraint. For the rest of the process, think VERY LARGE equipment. Tens of millions of dollars in equipment are required for a ready-to-eat cereal production line.

Putting the Ingredients Together

Finally! We have the ingredients and are ready to make the finished cereal. After the other processes that we've already reviewed, this may seem almost anticlimactic.

Shredding the Wheat

Raw wheat berries are received at the cereal production facility, usually by rail car. Each rail car brings in approximately 215,000 pounds of wheat, which is blown into silos by air. Depending on the product's nature and the desired outcome, several ways of moving it are available.

1. The wheat berries are sieved and/or washed to remove any remaining stones, debris, or even undersized berries.
2. The berries are transferred to either a continuous or batch-type cooker. In a continuous cooker, the wheat is transported by means of a screw through a large amount of water that's just short of the boiling point. In a batch cooker, the wheat is added to a vessel similar to a pressure cooker, along with a specific amount (not an excess) of water.
3. The cooked wheat is rinsed to remove the surface starch that has seeped out of the wheat berries.
4. Surface moisture is removed by traveling through a dryer.
5. The cooked, surface-dried wheat is allowed to sit for a specified period of time in order to let the berries equili-

> ***Cookers, Batch vs. Continous:*** In *continuous cookers*, raw material enters at one end and is conveyed continuously from the entrance to the exit. There, the now cooked product leaves the cooker. *Batch cookers* utilize several steps. A good analogy is a pot on a stove. The material you're cooking is added to the cooker and the lid is closed. This represents a batch, sometimes referred to as a *cook*. The batch is cooked in that that closed environment, then it's dumped out and the process starts over again, one after another.

brate, or come to the same moisture content throughout the berry. This allows the cooked starch to retrograde, or for the starch molecules to realign in an orderly fashion. Without this tempering time, it wouldn't be possible to shred the wheat. If you were to try to shred wheat that wasn't properly tempered, it would smear on the mill rolls. That would cause a huge mess and a resulting loss of production time!

> An *excess* means a lot—more than seems necessary. A good example is boiling pasta, which is done in an excess of water. You wouldn't add an excess of water to make a flour dough, though. You'd add a measured amount.

6. After the cooked wheat is tempered, it's transferred to shredding mills that consist of two rollers that press against each other. One roller is smooth, while the other has grooves in it. The grooved roller also has a stationary comb that rests in the grooves. Cooked, tempered wheat is pulled into the rotating mills and the comb peels the shredded wheat out. The shredded wheat now looks like long, thin, flexible strands that are similar to spaghetti.
7. The long thin strands are layered on top of each other to achieve the desired pattern and number of layers for the product. It creates a large bed—almost like a soft carpet—of shredded wheat.
8. At some point in this process, the powdered mix of dry vitamins is added. Depending on the manufacturer, it might be mixed in with the cooked wheat prior to shredding, or it can be sprinkled onto the shredded wheat bed between the layers.
9. The finished bed, or matt of shredded wheat, is crimped and cut to achieve the desired biscuit shape and size. At this point, we have a bed of slightly connected biscuits that are the correct size, shape, and thickness, and are ready to be baked.
10. The bed of biscuits is fed into a very long, continuous convection oven. These ovens can be the length of a

football field, and sometimes even longer. The temperature and airflow are set to result in the correct amount of toasting color and flavor to the biscuits, and also to dry them to the correct moisture content.

11. For a non-frosted biscuit, the product is finished! Basic toasted shredded wheat is about as simple a product as you can get. It's boiled, smashed, shredded, and toasted wheat.

Frosting

Frosting can be produced in several ways. It can start with different forms of sugar, but in this case you'll see a process that starts with granular sugar, where the frosting is made in batches (see above).

1. Due to the large quantities needed, the sugar is also received in rail cars. Each car weighs approximately 220,000 pounds. The cars are emptied and the sugar blown by air into silos, where it's eventually brought to the production line.
2. The granular sugar's crystal size is too large to be made into icing directly. So it's ground down to powdered sugar using various types of pulverizing mills. These mills can have rotating discs that grind or hammers that hit and pulverize, to name a couple of methods.
3. Water and gelatin are added together, heated, and allowed to sit for a certain amount of time. This allows the gelatin to hydrate, or take up water, and eventually disperse into the water. The *function* of gelatin in the finished frosting is to provide durability and to prevent the frosting from dissolving too quickly when milk is poured onto the finished cereal.

Hammer mill: https://www.youtube.com/watch?v=e6trUtolOZE, https://www.youtube.com/watch?v=Y93rf2cNlol

Function or Functionality: *These words describe what a food component does in a situation. The functionality of a component can be changed, or even lost, depending on how it's treated. For example, egg whites function well to make foams (such as meringue), but that functionality is destroyed when heated.*

4. The freshly powdered sugar is put into a mixing kettle, along with water and gelatin, and they're blended together. At this point, the icing is finished and ready for use.
5. The icing is sprayed onto the still-warm biscuits, using different types of spray equipment depending on the desired appearance and pattern.
6. The icing must dry and set before the bed of freshly frosted biscuits is disturbed. After that happens, they're ready for packaging.

There you go! Shredded wheat is a very simple product, and adding frosting to it doesn't complicate things too much. However, it takes quite a while to explain how it's produced, and a considerable amount of very large and expensive equipment.

Part of the purpose in going through this exercise is to reiterate that shredded wheat is a processed food. Anything you do to make a raw food edible is a process! You shouldn't be afraid of processed food, but with your new knowledge you can choose wisely…and impress your friends!

Making Shredded Wheat
https://www.youtube.com/watch?v=oWFFXuhvkKA

WHAT DO YOU THINK?

1. **Give five examples of a minimally processed food that you eat at home.**
2. **What are the five most highly processed foods currently in your house?**
3. **Why do you think that the term** *processed food* **has such a negative connotation?**
4. **Now that you know more about processed food, do you think it deserves the negative view that it has? Why or why not?**
5. **How can you educate yourself about the processed food debate's pros and cons? Talk to someone about what you think and see if they agree.**

JOURNALING IDEA

Now that you have a better understanding of what minimally and highly processed foods are, pretend that you're currently 70 years old and you've lived your life eating mostly highly processed foods. Describe what you look like, and how you feel.

Next, pretend that you're 70 years old again, but this time you've lived your life eating mostly minimally processed foods or foods made at home, only consuming highly processed foods when you needed the convenience or wanted a treat. Describe what you look like, and how you feel.

Which person do you want to become? Write down some goals about how you want to eat and how you'd like to feel as you grow older. Discuss these goals with someone close to you.

CHAPTER REVIEW

Write short essay responses to each of the following questions:

1. Name a food that you have processed before it was ready to consume and describe this process. Example: you may have washed, peeled, then cut a carrot into bite size pieces.
2. Describe in detail the most and least processed food you consumed over the last 5 days.

CHAPTER 4
PROXIMATE ANALYSIS

THE FIVE BASIC COMPONENTS OF ALL FOOD

All foods have five primary components: water, protein, fat, carbohydrates, and ash. Water and protein are self-explanatory. Fats include saturated, monounsaturated, and polyunsaturated fat and trans fat and cholesterol, among others. There will be more on fats in Chapter 8. Carbohydrates include starch, fiber, and sugars. Ash is everything else, and is composed primarily of minerals.

Breaking down food into these five basic components is referred to as a *proximate analysis,* or *composite analysis.*

MEASURING WATER, CARBOHYDRATES, PROTEIN, FAT, AND ASH

While a thorough description of laboratory procedures used to measure the basic food components is beyond the scope of this workbook, a short description for each will help.

Water

The amount of water is determined by weighing the food very accurately, drying it, and then weighing it again. The difference between the two is the amount of water the original sample contained. While this sounds simple, it can become complex

Water[34]

as some foods, such as a gummy bear, don't dry well without additional procedures.

A simple formula:

(Wet Weight) - (Dry Weight) = Water Weight

Ash

Ash is measured by burning a food sample completely and weighing the ashes that are left. Another term for burning material in this manner is *incineration.* A very small amount of ash is left after incineration, and it consists primarily of minerals.

Ash[35]

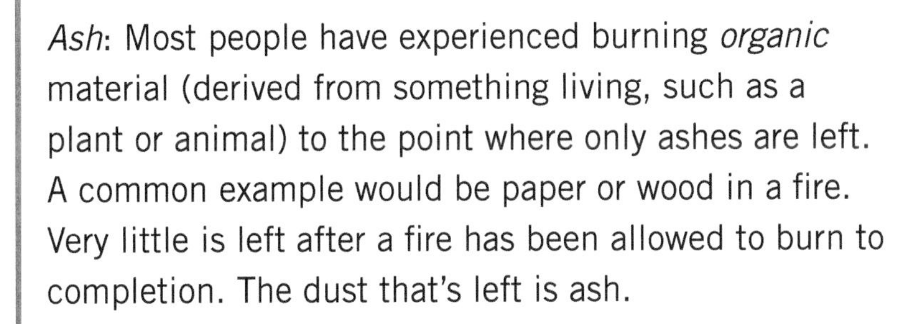

Ash: Most people have experienced burning *organic* material (derived from something living, such as a plant or animal) to the point where only ashes are left. A common example would be paper or wood in a fire. Very little is left after a fire has been allowed to burn to completion. The dust that's left is ash.

Carbohydrates

Measuring the many different carbohydrate types is tricky. Because of this, usually the total amount of carbohydrates is determined by a difference calculation. That is, add up everything else and subtract that sum from the total to get the amount of carbohydrates:

Total Weight - (Sum of Protein + Fat + Moisture + Ash) =
Carbohydrates Weight

Determining what type of carbohydrate the amount is will be much more difficult. For example, glucose, sucrose, and starch (you'll learn more about these later!) are common carbohydrates. Analyzing a food sample and determining which are present and in what amounts is very challenging. In this workbook, suffice it to say that with the right equipment and laboratory methods, the amount of and types of fiber and sugar can be very accurately measured.

Protein

All protein contains the element nitrogen in predictable quantities, and the element is almost exclusively found in protein. By measuring the amount of nitrogen, the amount of protein in a sample can be calculated. A procedure referred to as the Kjeldahl Method is one way to do this, and involves digesting (the actual word used in the method) a food sample with sulfuric acid. The sulfuric acid reacts with nitrogen in the food to produce a byproduct, and the amount of that byproduct is measured by *titration*.

Fat

Butter[36]

Fat can be measured in foods also by various methods. One method is by running a fat solvent, such as ethanol, over the food sample until all fat has been removed. The fat/ethanol mixture is heated to allow the ethanol to evaporate, leaving just the fat, which is weighed. This is a fun experiment to do. You just have to be careful not to blow yourself up, as the fumes are flammable!

Titration: In very simple terms, titrating is adding a solution that reacts with and uses up a component in the sample liquid in a predictable fashion, with an indicator for when that component is all gone. This indicator can be a color change. For example, Solution B is prepared from the sample through various methods, including, in this case, digestion. Solution A is slowly added to Solution B until all the material is gone, at which point the combined liquid's color changes. Knowing the amount of Solution A that was used allows the quantity of components in Solution B to be calculated.

THE NUTRITION FACTS FOOD LABEL

The U.S. Nutritional and Labeling Education Act of 1990 introduced major changes to the side panel labels on food packages. Manufacturers and producers were required to inform consumers what was in the package of food. For instance, the side panel for crackers contains a lot more ingredients than just flour, salt, water, oil or butter, and some flavorings. Go take a look in your cabinet!

In May of 2016, the Food and Drug Administration announced enhancements to the label, which companies have to comply with fully by 2021. The photo below shows the differences between the two. As of 2019, both are in use.

SIDE-BY-SIDE COMPARISON

Original Label

Nutrition Facts
Serving Size 2/3 cup (55g)
Servings Per Container About 8

Amount Per Serving	
Calories 230	Calories from Fat 72
	% Daily Value*
Total Fat 8g	12%
Saturated Fat 1g	5%
Trans Fat 0g	
Cholesterol 0mg	0%
Sodium 160mg	7%
Total Carbohydrate 37g	12%
Dietary Fiber 4g	16%
Sugars 12g	
Protein 3g	
Vitamin A	10%
Vitamin C	8%
Calcium	20%
Iron	45%

* Percent Daily Values are based on a 2,000 calorie diet. Your daily value may be higher or lower depending on your calorie needs.

	Calories:	2,000	2,500
Total Fat	Less than	65g	80g
Sat Fat	Less than	20g	25g
Cholesterol	Less than	300mg	300mg
Sodium	Less than	2,400mg	2,400mg
Total Carbohydrate		300g	375g
Dietary Fiber		25g	30g

New Label

Nutrition Facts
8 servings per container
Serving size 2/3 cup (55g)

Amount per serving	
Calories	**230**
	% Daily Value*
Total Fat 8g	10%
Saturated Fat 1g	5%
Trans Fat 0g	
Cholesterol 0mg	0%
Sodium 160mg	7%
Total Carbohydrate 37g	13%
Dietary Fiber 4g	14%
Total Sugars 12g	
Includes 10g Added Sugars	20%
Protein 3g	
Vitamin D 2mcg	10%
Calcium 260mg	20%
Iron 8mg	45%
Potassium 235mg	6%

* The % Daily Value (DV) tells you how much a nutrient in a serving of food contributes to a daily diet. 2,000 calories a day is used for general nutrition advice.

Nutrition Facts, new and old[37]

Using the Nutrition Facts side panel, it's possible to determine the proximate analysis of any packaged food. The side panel for Kellogg's Raisin Bran® shows the grams, or milligrams, of the different components. Remember, the global food industry uses metric measurements.

The table below has been filled out for Raisin Bran from the pictured Nutrition Facts panel.

Calculated Proximate Analysis

	Units Used On Nutrition Facts Side Panel (g or mg)	From Nutrition Facts	Convert to grams (if needed)	Calculated Percentage
Total Fat	g	1.0	1.0	1.7%
Total Carbohydrate	g	46.0	46.0	78.0%
Ash (add mineral values together)	mg	600.0	0.6	1.0%
Protein	g	5.0	1.0	1.7%
Water: (serving size) minus (fat+carbohydrate+ash+protein)			10.4	17.6%
		Total Percentage		100.0%

Proximate analysis chart, Raisin Bran[39]

Raisin Bran side panel[38]

Note that milligrams is the unit of measure in the Nutrition Facts panel for sodium and potassium. In order to complete an accurate proximate analysis, everything has to be in the same units. So milligrams (mg) will need to be converted to grams (g) by dividing by 1,000.

For example, in order to calculate the total ash—consisting of minerals (in this case, sodium and potassium)—add up the minerals listed in milligrams and divide by 1,000, since 1 gram (g) equals 1,000 milligrams (mg):

$$(210\ mg + 390\ mg) \div 1{,}000\ mg\ per\ g = 0.6\ g\ ash$$

Below is the same table, but showing the math. You'll want to take note of it for the assignment at the end of this chapter.

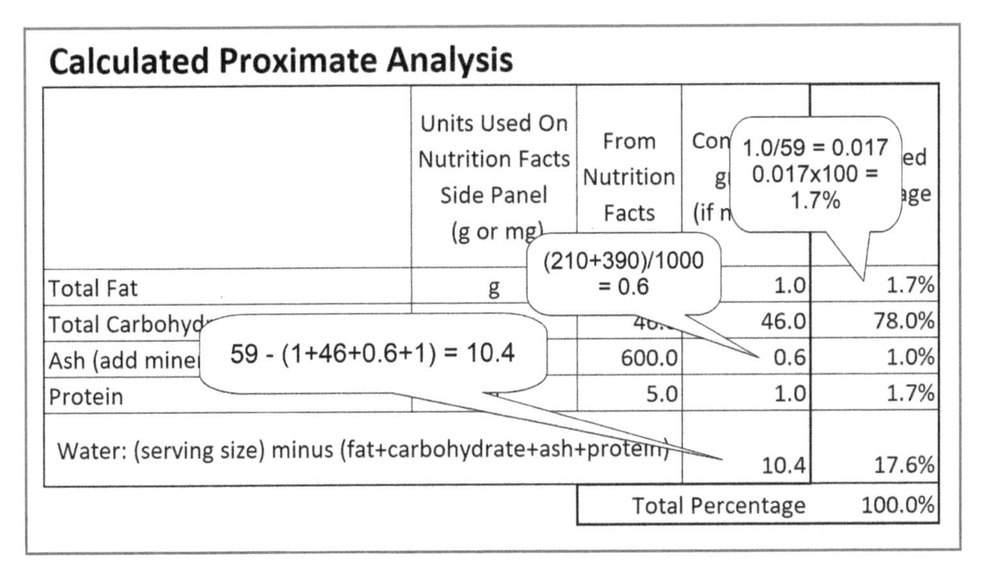

Proximate analysis chart, Raisin Bran, showing math[40]

When evaluating the Raisin Bran results, you can see that the product is primarily carbohydrate, with some protein and a little fat. The specific sources of carbohydrate show quite a bit of sugar (18 g), a lot of fiber (7 g), and 21 g of other carbohydrates[xii] that consist of longer chained, digestible carbohydrates. (Don't worry; you'll learn more about these later.) The Nutrition Facts panel is very useful for determining what's in your food!

Knowing what's in a food can also be very useful for the food scientist. For example, a common task is to make a product similar to what already exists in the marketplace. A quick analysis of the side panel, along with public information from suppliers about the individual ingredients and the associated ingredient nutrition information, as well as some experience picking apart competitive food products, can get you close to the existing food product's formula. Rather than completely reinventing the product, this process can be a good starting point for further product research and development!

Now that we know the basic major components of food, we can evaluate each component in more detail in subsequent chapters.

xii [(46 g total - 18 g sugar - 7 g fiber) = 21 g] other carbohydrates.

WHAT DO YOU THINK?

Look at the Nutrition Facts panel for five of your favorite foods and complete the Calculated Proximate Analysis tables on the following pages. Try to choose foods that are different from each other—for example, a breakfast cereal, mixed frozen vegetables, a candy bar, a carbonated beverage, and hot dogs. It's likely that some foods you choose may not have all the items that are listed on the worksheet. There may also be foods that have more than what's listed. Manufacturers have some flexibility on the label, and sometimes want to accentuate that a food does contain a certain nutritional component. Do the best you can.

After you've completed filling out the forms, answer the questions below.

1. How do the foods differ from one another?
2. Think about what you eat in a typical day. Are you mostly getting a lot of one component, such as carbohydrate or fat? The diet of a typical U.S. citizen is high in refined carbohydrates such as sugar or finely milled flours.
3. Pretend that today is your first day on the job as a food scientist and you've been asked to design a food which is similar to one of those you've just analyzed. What might be your next step?

Calculated Proximate Analysis Worksheet #1

Product	
Serving Size (grams – g)	
Total Fat	
Saturated Fat	
Trans Fat	
Polyunsaturated Fat	
Monounsaturated Fat	
Cholesterol (mg)	
Sodium (mg)	
Potassium (mg)	
Total Carbohydrate (g)	
Dietary fiber	
Sugars	
Protein (g)	

Calculated Water

	grams (g)	percent (%)
Total Solids: (Sum of Fat, Carbohydrate, and Protein + Ash		
Water (Serving Size minus Total Solids)		
Total		

Calculated Proximate Analysis

	grams (g)	percent (%)
Total Fat		
Total Carbohydrate		
Protein		
Water		
Ash		
Total		

Calculated Proximate Analysis Worksheet #2

Product	
Serving Size (grams – g)	
Total Fat	
Saturated Fat	
Trans Fat	
Polyunsaturated Fat	
Monounsaturated Fat	
Cholesterol (mg)	
Sodium (mg)	
Potassium (mg)	
Total Carbohydrate (g)	
Dietary fiber	
Sugars	
Protein (g)	

Calculated Water

	grams (g)	percent (%)
Total Solids: (Sum of Fat, Carbohydrate, and Protein + Ash		
Water (Serving Size minus Total Solids)		
Total		

Calculated Proximate Analysis

	grams (g)	percent (%)
Total Fat		
Total Carbohydrate		
Protein		
Water		
Ash		
Total		

Calculated Proximate Analysis Worksheet #3

Product	
Serving Size (grams – g)	
Total Fat	
Saturated Fat	
Trans Fat	
Polyunsaturated Fat	
Monounsaturated Fat	
Cholesterol (mg)	
Sodium (mg)	
Potassium (mg)	
Total Carbohydrate (g)	
Dietary fiber	
Sugars	
Protein (g)	

Calculated Water

	grams (g)	percent (%)
Total Solids: (Sum of Fat, Carbohydrate, and Protein + Ash		
Water (Serving Size minus Total Solids)		
Total		

Calculated Proximate Analysis

	grams (g)	percent (%)
Total Fat		
Total Carbohydrate		
Protein		
Water		
Ash		
Total		

Calculated Proximate Analysis Worksheet #4

Product	
Serving Size (grams – g)	
Total Fat	
Saturated Fat	
Trans Fat	
Polyunsaturated Fat	
Monounsaturated Fat	
Cholesterol (mg)	
Sodium (mg)	
Potassium (mg)	
Total Carbohydrate (g)	
Dietary fiber	
Sugars	
Protein (g)	

Calculated Water

	grams (g)	percent (%)
Total Solids: (Sum of Fat, Carbohydrate, and Protein + Ash		
Water (Serving Size minus Total Solids)		
Total		

Calculated Proximate Analysis

	grams (g)	percent (%)
Total Fat		
Total Carbohydrate		
Protein		
Water		
Ash		
Total		

Calculated Proximate Analysis Worksheet #5

Product	
Serving Size (grams – g)	
Total Fat	
Saturated Fat	
Trans Fat	
Polyunsaturated Fat	
Monounsaturated Fat	
Cholesterol (mg)	
Sodium (mg)	
Potassium (mg)	
Total Carbohydrate (g)	
Dietary fiber	
Sugars	
Protein (g)	

Calculated Water

	grams (g)	percent (%)
Total Solids: (Sum of Fat, Carbohydrate, and Protein + Ash		
Water (Serving Size minus Total Solids)		
Total		

Calculated Proximate Analysis

	grams (g)	percent (%)
Total Fat		
Total Carbohydrate		
Protein		
Water		
Ash		
Total		

JOURNALING IDEA

Ask an adult you know if their doctor recommends they avoid certain foods, or components of a food, such as sodium. Write about how using the Nutrition Facts side panel can help with this task. What about foods prepared at home? Note: Some questions in this workbook, including ones below, require thinking beyond the material presented in the workbook. Be creative and use your mind...you can do it!

CHAPTER REVIEW

Write short essay responses to each of the following questions:

1. How can the Nutrition Facts found on labels help you plan your diet? Do a couple of searches on the internet and write about what information used to be on labels before the Nutrition Labeling and Education Act of 1990.
2. What would you do if packaged food didn't contain nutrition information?
3. Write about the school experience of an adult you know. Did they go to college? If so, did they graduate in the major they chose when they began? Are they still working in that same field? Write about what this may mean for your own life.

CHAPTER 5
WATER

Water is a crucial compound for life, as well as for food. While this might seem clear, some underlying processes associated with water aren't widely known. For example, water is a great solvent, which makes it useful for cleaning and also for dissolving other materials. In food, almost no other ingredient has as much control over a product's texture. Too much water in a food may make it too soft. The opposite is also true: too little water and the food may become overly brittle or hard. Water also plays a large role in how much time can pass before a food starts to develop bad flavors, or even becomes unhealthy due to microbial growth! Let's explore some of these concepts.

Glass of water[41]

BIPOLAR IN NATURE

As we discussed, a water molecule is composed of 2 atoms of hydrogen bonded covalently to 1 oxygen atom. The result is an unbalanced molecule with a negatively charged side and a positively charged side.

In this depiction of the water molecule, the hydrogen atoms are light in color and the oxygen atom is dark.

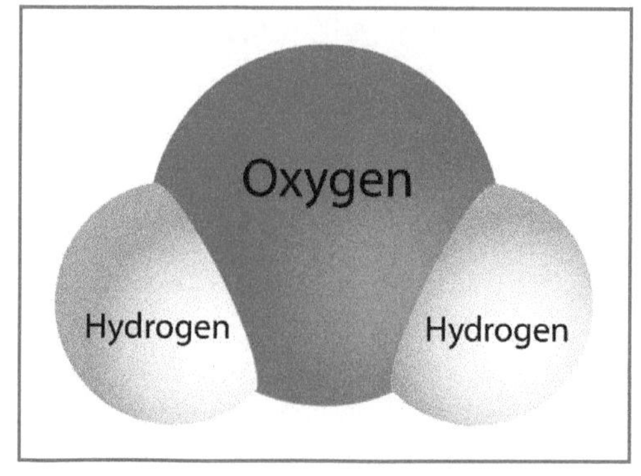

Water molecule[42]

Note how in the spatial orientation, the hydrogen atoms are on one side. This side has a positive charge, while the other side has a negative charge.

The resulting opposite charges make the molecule bipolar in nature. This appears to be a large part of the reason why it's such an important component in everything that has to do with life, including making food. Due to its bipolar nature, the water molecule is able to act as a bridge between other negatively and positively charged molecules. The water molecule is also very small, which allows it to work its way into and between most compounds.

For these reasons, the water molecule is a very good solvent. Have you ever noticed how something dirty becomes clean relatively easily when you include water? When washing dishes, you may have learned that when dealing with something really baked on, letting it soak for a while in water will often allow whatever was once like concrete to become soft, pliable, and easy to remove, and may even completely dissolve. This is an example of water's solvent effect. Something to think about the next time you wash the dishes!

Water as Solvent Mini-Experiment

Place 2 drops of honey or pancake syrup on your hand and try to wipe it off with a paper towel. Were you successful with just the paper towel? Now put about a tablespoon of water on another paper towel and use the wet portion to clean your hand. What did you notice?

WATER ACTIVITY: a_w

An extremely important concept in food science is a measure of how *available* water is from an ingredient or food. One measure of this property is a concept called *water activity*, which is different than simply how much water is in the product. This measurement demonstrates how tightly a food is holding the water. Water activity, represented by a_w, is defined as "the partial pressure of water as measured above a food or ingredient, divided by the partial pressure of water as measured above pure water when both the food and water are at the same temperature and in a closed container, or *system*."

For example, a common design for an instrument that measures a_w has a small chamber where a food sample is placed and a tightly sealing lid is closed. After the system equilibrates, the water pressure is measured in the space above the food sample. In equation form, it looks like this:

$$a_w \text{ (Water Activity)} = \text{Partial Pressure of Water Over the Food } (p^o) \div \text{Partial Pressure of Water Over Pure Water } (p^w)$$

Or, without the explanation:

$$a_w = (p^o) \div (p^w)$$

Another way to think about a_w is in terms of relative humidity, which is more commonly understood and something that people can relate to:

$$a_w = p^o \div p^w = \text{Equilibrium Relative Humidity (ERH)} \div 100$$

In geographic locations where the outside relative humidity is high for much of the year, such as the United States' southeastern region, a much larger mold problem exists than in states where humid days aren't as frequent. Take North Carolina and Colorado as examples—both states in which I've lived. In North Carolina, algae grows on the side of a house easily. It's a normal process to have to scrub off house siding every 2–3 years. If you wait longer, you might even have snail trails crawling around on the side of your house as they consume the algae!

In Colorado, where the relative humidity is much lower, very rarely is this a problem. In this example, the water in the air at the siding's surface is sufficient to promote algae growth in North Carolina, but not in Colorado, where there is very little water in the air.

In the same way, think of a fresh piece of bread right after you take it out of the bag. It's soft, moist, and flexible. While it isn't visible, water is available *at the surface* for interaction.

> ***Water "At the Surface":*** This phrase might seem a bit confusing, but is a way to think about water in a food that is important. Water is mixed throughout a food, but some reactions require the water on the surface to be available enough to begin.

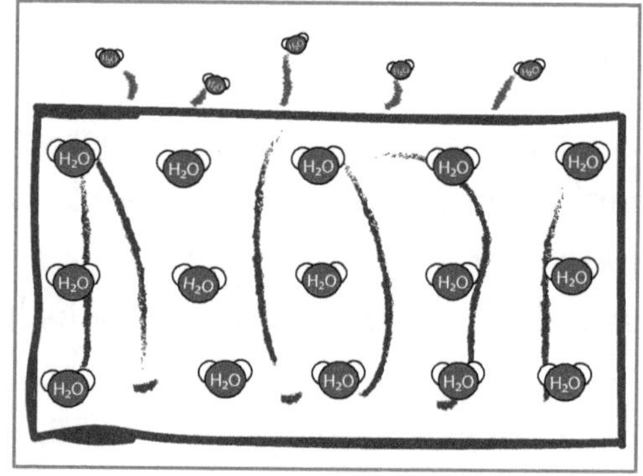

Freshly sliced bread[43]

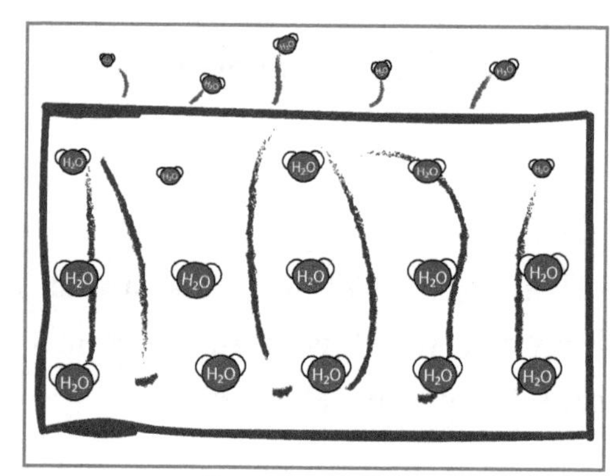

Partially dried bread[44]

Now consider a cracker. It's crisp and fragile, showing us that very little water is available for interaction with other things, such as bacteria or mold. The bread's a_w is higher than the cracker's a_w. Or you could say that the relative humidity at the bread's surface is higher than the relative humidity at the cracker's surface.

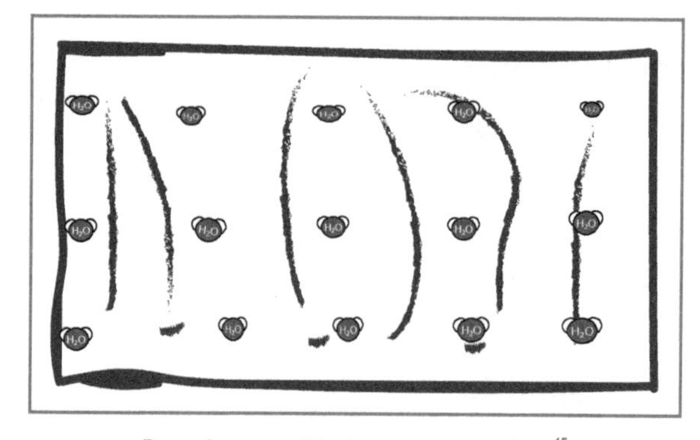

Bread at equilibrium, or a cracker[45]

This measurement of water's availability can be used to determine where there may be issues with bacterial or mold growth. It's also useful knowledge when combining ingredients and predicting the relative transfer of moisture between those ingredients.

It's important to match the a_w of all components when they're going to be together in an enclosed package! If this can't be done, at least the equilibrated product's characteristics need to be acceptable. *Fully equilibrated components* mean that all available water has been lost or gained by the ingredients and reached a point of equilibrium, or a steady state. Using our earlier example, if you put the piece of fresh bread and the cracker in an enclosed container, the cracker will absorb moisture from the bread and become less crisp and more flexible, and the bread will become less soft and less flexible. This equilibration process begins immediately. The effects on the cracker will become noticeable in a very short period of a time, even minutes to an hour, but full equilibration might take a day or so and depends on conditions like the container's size, the air's humidity when the container was closed, and the bread and cracker's initial moisture contents.

WATER DENSITY

Water is unique in that it's the only molecule that exists in all three phases naturally at temperatures on the Earth's surface. Liquid water, solid ice, and water vapor are all routinely found together when the conditions are right. Some water vapor is always in the air, even in areas where the humidity is extremely low, such as in a desert environment. In cold locations where it snows, when the temperature is near the freezing point, both solid and liquid water will exist, along with water vapor in the air.

Iceberg floating in the ocean[46]

A strange characteristic of water is that when it changes to crystal form, such as snow and ice, it becomes less dense. A reduction in density is the opposite of what's observed everywhere else in nature. Other substances become more compact, and therefore denser, when crystallization occurs. Think about lava, for example. You won't see solid rocks floating on top of a lake of lava. Lava itself actually "floats" to the top since it's less dense than the surrounding solid rock. For this to occur with water, a glass of water and ice would have to contain the ice cubes on the bottom. Of course, this doesn't happen. Instead, ice floats in water!

THE FORMATION OF ICE

The formation of ice, or the crystallization of water, is a complex process. The fact that ice is less dense than water is primarily due to the hydrogen bonds that exist between the molecules of water. These bonds form between a hydrogen atom of 1 water molecule and the oxygen atom of a neighboring water molecule.

Hydrogen bonds are usually considered weak, but when together in large numbers, they can create quite an impression. Just ask anyone who has fallen on their rear end while ice skating. Ice is very solid stuff!

Hydrogen bonds are also largely responsible for the fact that water holds together in liquid form. Without these bonds, water might just all float away as vapor. Another example of hydrogen bond strength is when you do a belly flop into a lake or a pool. It really hurts! The water holds together and doesn't move without some resistance.

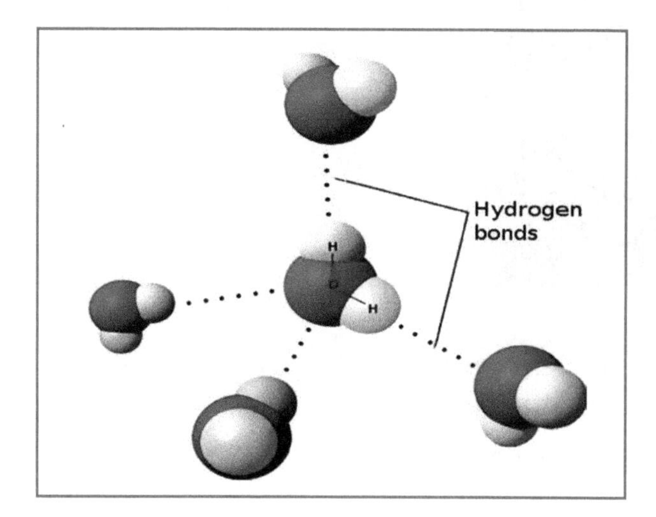

Hydrogen bonds[47]

In the illustration above, dotted lines represent the hydrogen bonds between the oxygen atoms of 1 water molecule and the hydrogen atoms of neighboring water molecules.

Ice formation and hydrogen bonds: https://www.youtube.com/watch?v=UukRgqzk-KE

IMPLICATIONS IN FOOD

Separation

When ice forms in a food system, it forces impurities, or things that aren't water, out of the crystalline structure. This concentrates the impurities, and chemical reactions will sometimes occur that otherwise wouldn't have. Some of these reactions may cause off-flavors, unexpected colors, or otherwise spoil the food.

Damage

In addition, as ice forms, we've already seen that it expands. This expansion can burst the cells of fresh plant foods, causing irreversible damage. Ice-induced expansion can also burst packaging, sometimes dangerously. For example, you should never freeze carbonated beverage containers. They can explode and potentially cause harm, and at the very least, make a mess. Below are a couple of video clips showing this phenomenon in action.

Exploding frozen carbonated beverage cans. https://www.youtube.com/watch?v=RGhgpJAHsT8 https://www.youtube.com/watch?v=WFyaL6iozKY

Freezer "Burn"

A major challenge in the frozen food industry is *frost-free* freezers. The current generation may not even know what it means, since it's no longer a real selling point for freezers. Almost all of them are frost-free.

Freezers pull moisture out of the air. The moisture enters when the door is opened, and then ice forms on the freezer's inside surfaces. This *frost* used to continue to build up over time and eventually took up a large portion of the space inside the freezer. Periodically, a major effort was required to remove the frost and recover the lost space. When the buildup became very thick, the freezer would also cease to function effectively.

Today's frost-free freezers go through a regular freeze/thaw cycle. The freezer warms sufficiently to melt any frost that has developed and channels it to drip into an exterior pan where it eventually evaporates. The thaw cycle is short and doesn't allow the food to thaw much, but it does thaw to some extent.

This cycle is extremely abusive to frozen foods. It causes a concentration of sugars, flavors, and oils, as well as surface dehydration. This concentration and accompanying growth of ice crystals change the texture. All of these together are sometimes referred to as *freezer burn* because the food surface may appear burned or disfigured. It requires a lot of time and research to create a frozen food that will last in a modern freezer for a significant period of time and will still taste good when it's consumed. It's one of the most difficult challenges that a food scientist in the frozen food industry will experience!

Speed is Important

Just as freeze/thaw cycles in a freezer will generate larger crystals, if a food is frozen slowly, ice crystals have an opportunity to grow larger. In ice cream, for example, *organoleptically* detectable ice crystals are extremely undesirable.

Therefore, ice cream is made in such a way so that freezing can occur while keeping the crystals small, and ingredients are added to help minimize ice crystal growth during freezer freeze/thaw cycles.

Organoleptic: Evaluated by the senses, including sight, taste, and feel. In this case, ice cream in the mouth should be creamy and smooth, not icy.

EXPERIMENTS: LET'S MAKE A MESS!

Water Experiment #1: Water Activity and Moisture Balance

Background

Early in this chapter we discussed the concept of water activity (a_w) and how it's an important concept for food scientists, and anyone interested in preserving their food, to understand. In this experiment, you'll learn one aspect of a_w and see how it changes the nature of a finished food.

Items Needed

- 4 cups corn or bran flakes (NOT raisin bran flakes)
- 3/4 cup raisins, freshly opened
- 3 quart- or gallon-sized zip-closure type bags
- 2 teaspoons water
- Permanent marker or pen

> A freshly opened container of raisins is needed, whether they're in a bag or otherwise. Once opened, the raisins will start to equilibrate with their environment (the room's air) and their moisture content will change. This change will be an additional variable in the experiment and might alter your results.

Procedures

1. Label one bag "Flakes With Water," one bag "Flakes Without Water," and one bag "Raisins." Write the date and time on each bag.

2. Place 2 cups corn or bran flakes in each flake bag. Sprinkle 2 teaspoons of water onto the 2 cups of flakes in the bag labeled “With Water.” It’s all right if some water gets on the bag—the flakes will absorb the moisture over time. Close the bag and shake it around a little to help with the distribution of the water. In a cereal process, water would most likely be applied by spraying it onto flakes in a rotating drum.

Note: Don't add any more water than the 2 teaspoons. Too much water might move the a_w of the flakes enough to where mold growth can occur, making them unsafe to evaluate later.

3. Place 1/4 cup raisins in each bag. You should now have two bags of flakes with raisins, and one bag of just raisins. The bag of flakes and raisins should have been sprinkled with water.
4. Seal the bags. Place in a dry, cool place.
5. Wait one week.
6. Open the bags and evaluate a few flakes and a couple raisins for texture, filling out the below table. Reseal the bags. You may also note any other differences that are observed. Fill out the table below for Week 1.

Week 1	Flakes with Water	Flakes without Water	Raisins
Flakes			
Raisins			

7. Wait another week, then fill out the table for Week 2.

Week 2	Flakes with Water	Flakes without Water	Raisins
Flakes			
Raisins			

What do you think was going on? Describe it in your own words.

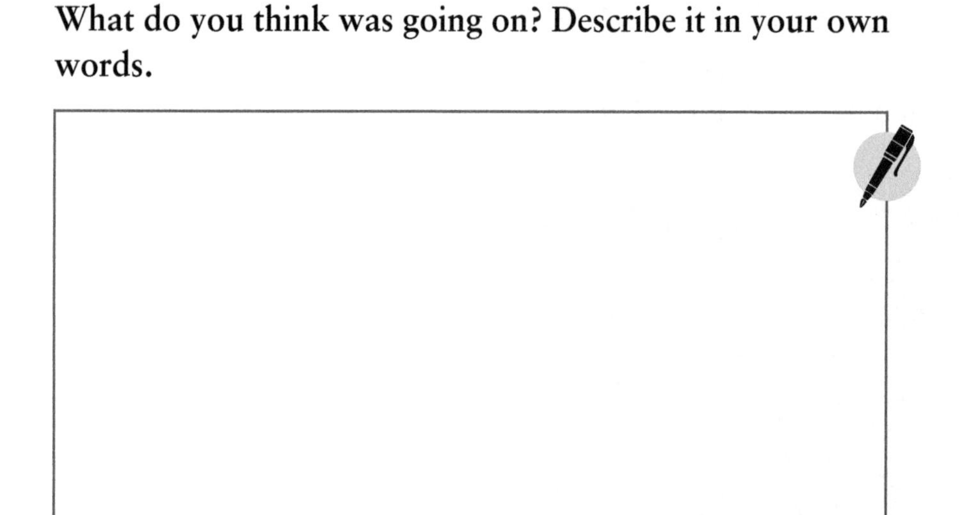

Water Experiment #1 Discussion: Water Activity and Moisture Balance

Through this experiment, you've seen how matching water activities is important to maintain or change characteristics in a food. When making raisin bran, manufacturers will apply water to the finished bran flakes to prevent them from pulling too much water out of the raisins, which would eventually make the raisins very hard. This process is called *equilibration* and will continue until an equilibrium is reached. That is, until no more moisture exchanges between the components. This is an important concept in food science and new product development!

Water Experiment #2: Crystal Density

Background

Water becomes less dense as it freezes, whereas most other pure substances become denser as they freeze and become solid. You'll be able to see this firsthand in this experiment.

Items Needed

- 2 clear containers of approximately 2 cups capacity each, tall enough so that at least 2 inches are visible on the container's side
- 1 small container that will hold 2 tablespoons of oil, about 1/4-inch deep
- Water ice—the size or shape doesn't matter, as long as it fits loosely in the larger container above
- Water, approximately 1 cup
- Olive oil, approximately 1 cup
- Masking or painter's tape
- Marker or pen

Note: By taking care and using clean containers, the olive oil can be poured back into the original container, or otherwise used for consumption after this experiment. It's expensive, so you don't want to waste it. Don't substitute another type of oil.

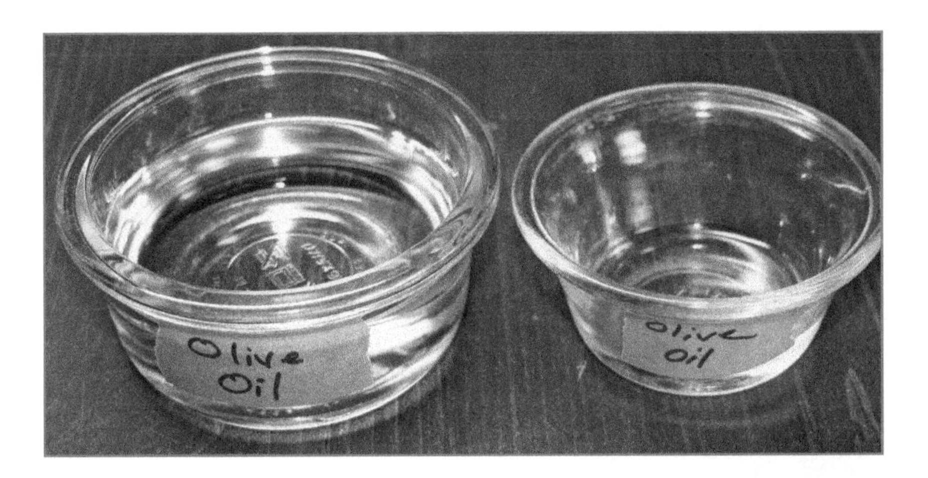

Procedures

1. Label one large container "Olive Oil" and the other "Water" by writing on a short strip of masking tape and sticking it on the container.
2. Add olive oil to the small container until it's about 1/4-inch deep, and carefully place it in the freezer. If you have a lid for this one, you can use it to help prevent spills. Otherwise, it isn't necessary. Let the oil freeze for 2 hours. It'll be opaque and hard when it's ready.
3. Fill the olive oil container with enough olive oil so that it's about 2 inches deep.
4. Fill the water container in the same manner as the olive oil container, until it's also about 2 inches deep.
5. Remove the olive oil from the freezer.

6. Immediately remove some solid, frozen olive oil with a spoon and place it in the larger container of olive oil. This can be done by scraping some out with a spoon. It doesn't have to be pretty, but there should be at least some large clumps of frozen oil.

7. Record your observations:

   a. What happened to the oil "ice?"
   b. Did it float or sink?

8. Repeat Step 7 using water ice and the larger water container.

9. Record your observations:

   a. What happened to the water ice?
   b. Did it float or sink?
   c. What happened to it as it melted?

Water Experiment #2 Discussion: Crystal Density Experiment

In this experiment, we saw that oil ice increases in density when it's formed, whereas water ice decreases in density. This unique property of water is different to almost all other compounds on the planet. It's important to keep this in mind when formulating products, and it can also be useful in food production processes.

Note: It's possible to freeze water and oil without changing their density very much. Extremely rapid temperature reduction, such as with liquid nitrogen, will turn the product solid before crystals have a chance to form. The crystallization process is an important factor associated with the change in density. If it isn't kept at a very cold temperature, the tendency will be for crystals to form as the solid ages. As crystals form, the density will change!

WHAT DO YOU THINK?

We've only scratched the surface of all the interesting things we know about water and the role it plays in processing food. The Edible Knowledge® workbook *Introduction to Food Science: Water* discusses this topic in much greater detail. Can you think of any food that doesn't use water, or have water in it to least some degree?

JOURNALING IDEA

Imagine what life would be like without water. Write a short science fiction story about a new colony on Mars and how the colonists had to recycle their water. Include methods and invent names for them, and specifically explain how you'd clean the water that's reclaimed from human waste, showers, and washing activities. Write about how you might be able to capture water that escapes through evaporation when preparing food.

CHAPTER REVIEW

Write short essay responses:

1. A building footing, or the solid foundation on which a building is constructed, must be deeper below ground in cold climates than in warmer climates. Why?
2. Your new food company, ABC Foods, would like to sell frozen ice cream that has organoleptically undetectable ice crystals. What might be your biggest hurdle, and why?
3. Think of the items around you that are solid, including the plastic in the pen you may be holding, the graphite in the pencil, or the plastic in the keyboard you may be using. Are these components frozen? Explain.
4. When a large wave approaches the shoreline of an ocean, the water immediately in front of it will pull back, or recede, into the ocean. This phenomena is noted distinctly when a tsunami (tidal wave) occurs. From what you've learned, you should be able to explain one of the primary causes.

CHAPTER 6
CARBOHYDRATES

Carbohydrates[48]

What are carbohydrates? Carbohydrates include a wide range of substances that can be anything from a simple sugar, such as glucose, to a complex polysaccharide, such as fiber. They are found everywhere in nature and compose much of the structure in the physical world of what we see on a daily basis. They make up the bulk of grasses, trees, and flowers, and fruits, vegetables, grains of all types, and sugars primarily consist of carbohydrates. They provide the bulk of the energy in a typical human diet.

Carbohydrate molecular structure can be straight-chained or have branches, and they can be small or very large molecules.

We'll explore some basics of carbohydrate chemistry and use some interesting experiments to demonstrate these concepts.

CARBOHYDRATE STRUCTURE

The most basic definition of a carbohydrate is "a molecule that consists of carbon, hydrogen, and oxygen atoms, usually with the hydrogen and oxygen ratio in a 2:1 relationship: 2 molecules of hydrogen to 1 molecule of oxygen."

Depending on how they are joined, carbohydrate structures have many names.

Sugars

Sugar (Sucrose) crystals[49]

Let's take a look at the differences between carbohydrate sugars.

Fructose, also known as levulose, is an example of a ketone. All ketones have 1 carbon molecule double-bonded to an oxygen molecule, with this structure located in the molecule's middle.

Glucose, also known as dextrose, is an example of an aldehyde. All aldehydes have 1 carbon molecule double-bonded to an oxygen molecule, with this structure located at the end of the molecule. (Don't worry too much about this terminology; we won't be getting into the chemistry very deep).

As with everything else in nature, sugar molecules are three dimensional and exist in many different orientations at the same time, including straight chains and ring structures. Nevertheless, to study and talk about them, we need to write them down in formats accepted by others in science, which can be challenging. Fischer[xiii] and Haworth[xiv] Projections, or representations, are two widely accepted formats. The Fischer Projection is simpler and shows molecules in a straight line. The Haworth Projection shows the ring structure.

Glucose, Fischer Projection[50]

Fructose, Fischer Projection[51]

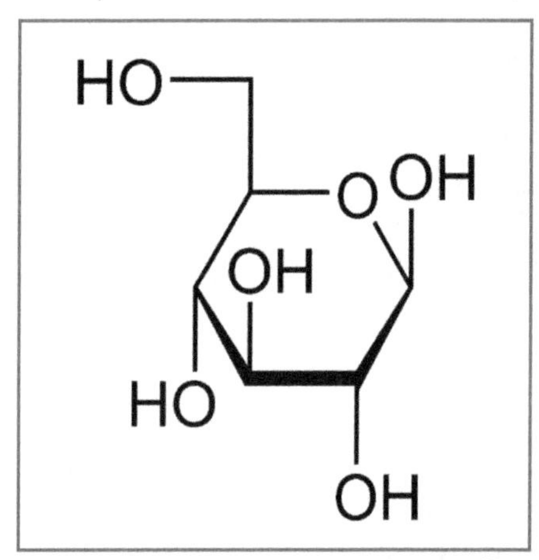

Glucose, Haworth Projection[52]

In nature, all configurations remain in balance relative to one another. The prevalence of each configuration depends on many things, including substance purity (Is it mixed with another thing?) and the environmental energy state (hot, cold, warm, etc.). Various configurations may chemically react differently with other molecules.

---

xiii This representation format was devised by the German chemist Hermann Emil Louis Fischer around 1890. The straight-lined, two dimensional format—as well as a 3D format not shown in this workbook—were so helpful that scientists commonly use them around the world today. Fischer was eventually awarded the 1902 Nobel Prize in chemistry for his combined work.

xiv This presentation format was named after a British chemist, Norman Haworth. Like the Fischer projections, these are widely used throughout the world today. Haworth also was awarded a Nobel Prize for his combined work, in 1937.

Starch

Glucose and fructose are two basic sugars that are very small. Another common carbohydrate is *starch*.

Textbox: *Starch*: You'll see this word many times through the workbook series. Starch is a generic term for the primary energy component of cereal grains and potatoes, among others.

The many types of cornstarch differ by plant species. Cornstarch can even be quite different depending on the type of corn. Context is also important, since cornstarch can be cooked or raw and has very different properties in these two states.

Starch molecules can be very large. For example, the average molecular weight for both glucose and fructose are approximately 180 grams per mole. The molecular weight for amylopectin, one of the two primary components of starch, can be anywhere from 360,000 to 36 million grams per mole.

In the next section, we'll examine some of sugar's characteristics. In a subsequent section, we'll evaluate some of starch's basic properties.

BASIC SUGAR PROPERTIES

Two of the more common sugars used in foods are sucrose, or table sugar, and fructose. Sucrose is a disaccharide, made up of 1 molecule each of glucose and fructose. Sucrose behaves quite differently than fructose or glucose, even though it consists of these 2 molecules joined together.

Non-Enzymatic Browning

Browning in food is of two primary classes: enzymatic and non-enzymatic. In this section we'll discuss non-enzymatic browning, also known as *Maillard Browning*, which requires the presence of certain types of sugars and protein, as well as heat. Both glucose and fructose can participate in these reactions. Non-enzymatic browning reactions are responsible for many flavors that we love in baked goods. This group of reactions is named after Louis-Camille Maillard, a French chemist who was the first to discover and describe the reactions. In addition to bread crusts, roasted meats and browned French fries are examples.

Monosaccharide: A simple sugar that includes glucose, fructose, galactose, etc. All monosaccharides are simple sugars of varying degrees of sweetness as perceived by the human tongue.

Disaccharide: 2 simple sugars bonded together. The best-known example is sucrose, which is common table sugar, and is made up of 1 glucose and 1 fructose molecule.

Polysaccharide: 3 or more simple sugars bonded together. A good example is cornstarch, which is essentially many glucose molecules bonded together into very long chains.

French bread[53]

The brown crust of bread and the flavors associated with it are good examples of non-enzymatic browning. The same occurs when making toast, although other reactions also occur.

Sucrose, which we know is what people are asking for when they want "sugar," is the most common sugar in our pantries. Its structure consists of a 1:1 combination of glucose and fructose, chemically joined in such a fashion that the molecule does not have the *reducing* structure that, for example, glucose has. Sucrose is as a *non-reducing sugar*, and is mentioned here because a reducing sugar is required for non-enzymatic browning reactions.

Maillard Browning:
https://www.youtube.com/watch?v=c7WI41huAok

An understanding of these types of reactions is essential for a food scientist. Sometimes these reactions are desired for a given food—and in fact necessary—to achieve the desired results. In other foods, their occurrence destroys the food. For example, dried powdered milk will turn brown over time in non-enzymatic reactions and will develop flavors that most people find offensive. In all cases, Maillard Browning needs to be controlled in order to achieve the desired finished product.

Sweetness

Sugars are very different in perceived sweetness. Commonly available sugars are fructose—the primary sugar in honey and corn syrup—and sucrose. Fructose tastes much sweeter to most people than sucrose. Understanding the sweetness level perception of different sugars is important for a food scientist. For example, if you want a food product with a particular sweetness level but don't want the side panel to show a lot of sugar, using fructose will result in more perceived sweetness than using sucrose. However, there can be other functional consequences, such as effects on browning and texture. A balance is usually required in order to achieve the desired result!

Non-Enzymatic (Maillard) Browning: https://www.youtube.com/watch?v=c7WI41huAok

Crystallization and Hygroscopicity

Another important aspect of sugars is their varying ability to crystallize, and to absorb or release moisture. For example, sucrose will crystallize easily compared to fructose, whereas fructose is more *hygroscopic* than sucrose.

STARCH CHARACTERISTICS

Amylose and Amylopectin

Starch has two primary components: amylose and amylopectin. Both consist entirely of glucose molecules strung together. However, amylose is straight-chained and amylopectin has branches.

Hygroscopicity: A measure of the attraction a substance has for water.

When dealing with starches, the structural difference between amylose and amylopectin is important. Corn and oats naturally contain amylose and amylopectin, but the amount of each is not the same. A measure that can be used to predict functionality is a ratio of amylose to amylopectin. Multiple species of corn available in the marketplace have a wide range of amylose:amylopectin ratios (read "amylose to amylopectin ratios"). Grains and starches are chosen for food applications based on this ratio so that the desired texture is achieved in the finished food.

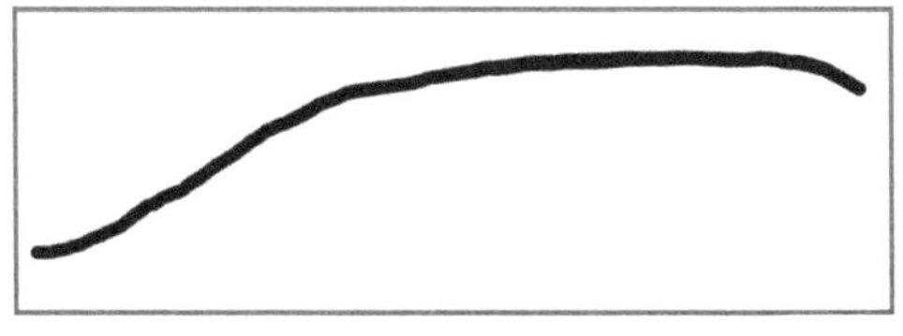

Amylose – Non-branched chains of glucose[54]

Amylopectin – Branched chains of glucose[55]

As an example, corn has been selectively hybridized to achieve many beneficial starch characteristics. Some varieties of corn contain a very high content of amylose, and some contain a high amount of amylopectin, which gives the basic vegetable corn diverse functional properties.

Gelatinization

Grains contain primarily starch. The amount and characteristics of the starch depend on the variety of grain and the growing conditions for a given crop, among other variables. Starches are laid down in *granules* as the plant grows, which are of different sizes depending on the variables just mentioned. The thickness of the layers is dependent on growing conditions. These include species and weather—temperature, exposure to sunlight, the amount of water, wind, etc. Also, the quality of the soil affects plant health. These considerations are all very important. It can be very frustrating when recipes don't turn out the same because of raw material variations!

The figures below provide a rough representation of cooking starch in water, a process which results in starch gelatinization. While the word *gelatinization* sounds like gelatin, they are not related at all, as you will see!

1. The starch granules are in an excess of water, loosely dispersed, prior to cooking.

Raw starch granules[56]

2. When starch is cooked in an excess of water, the granules begin to swell by absorbing water. As the granules continue to swell as water is taken up, they begin to bump into each other due to friction, as demonstrated by the thickening of the mixture.

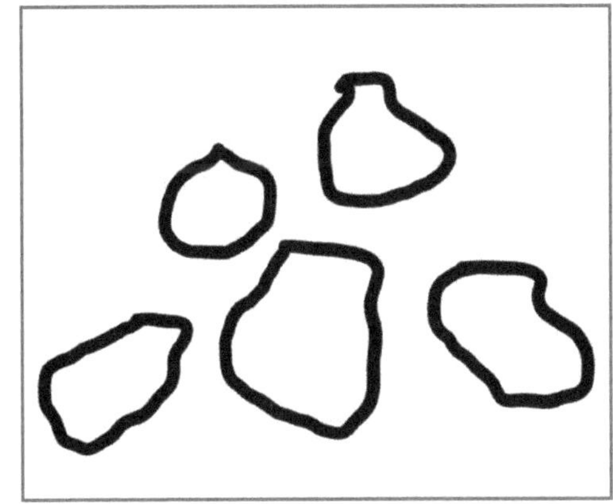

Partially gelatinized starch[57 58]

3. Cooking will continue in sufficient water, so the granules will almost completely swell and become very large, achieving maximum *viscosity* as the molecules continue to bump into and rub against each other.

Viscosity is a measure of thickness. Pudding is more viscous than water.

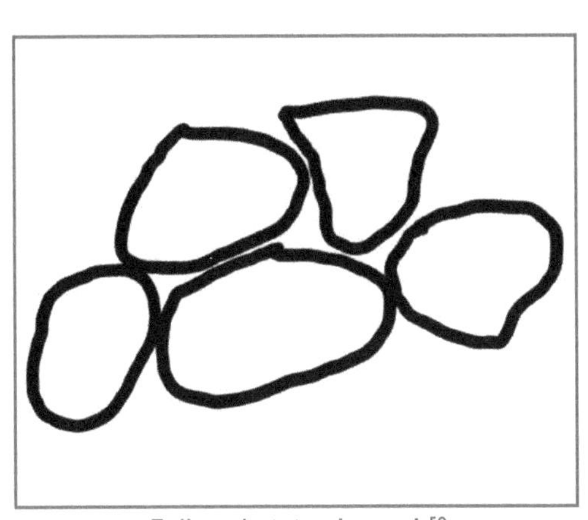

Fully gelatinized starch[59]

4. If the fully gelatinized, swollen starch granules are subjected to too much physical action at this stage, such as stirring, they may actually burst. This releases the hydrated starch chains into the general swollen starch dispersion—like the popping of a water balloon. When this happens, the viscosity is reduced. This is the reason why pudding instructions indicate to stir *gently*. Depending on the type of starch, there can also be a marked difference in consistency when this type of disruption occurs. Nobody wants pudding soup!

Starch granules swelling as they hydrate and gelatinize: https://www.youtube.com/watch?v=L6vYxYE1j0g

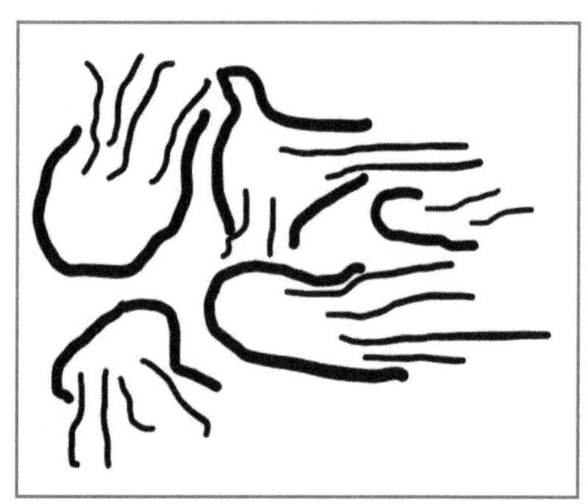

Broken starch granules[60]

EXPERIMENTS: LET'S MAKE A MESS!

Carbohydrates Experiment #1: Sucrose vs. Fructose

Background

In this exercise, we'll observe the differences in solubility and sweetness between fructose, sucrose, and corn syrup mixed into similar water contents.

Items Needed

- 1 tablespoon sucrose (normal table sugar)
- 1 tablespoon of dry fructose crystals (available in your grocery store, natural food store, or online)
- 1 tablespoon corn syrup (try to find some without other flavorings, such as salt and vanilla)
- Water, room temperature
- Teaspoon measurer
- Three small containers, such as shallow cups
- 3 teaspoons
- Masking or painter's tape
- Marker or pen

Procedures

1. Label the three containers using the tape and pen as follows: "Sucrose," "Fructose," and "Corn Syrup."

2. In the container labeled "Sucrose," dissolve 1 tablespoon of sucrose in 1 tablespoon of water by stirring with a spoon. Get as much crystalline sugar off the container's side and into the solution as possible. Leave the spoon in the container.
3. In the container labeled "Fructose," dissolve 1 tablespoon of fructose in 1 tablespoon of water by stirring with a spoon. As for fructose, get as much crystalline sugar off the container's side and into solution as possible. Leave the spoon in the container.

4. In the container labeled "Corn Syrup," add 1 tablespoon corn syrup, rinsing off the tablespoon with 1 teaspoon water into the container. Stir the corn syrup and water with the tablespoon measurer until the mixture is homogenous. Leave the spoon in the container.

5. Record your observations regarding how easily the crystals dissolved, or how the corn syrup's viscosity changed with the addition of water.
6. Taste each of the three mixtures in the following manner:
   a. Rinse your mouth with room temperature water, either swallowing or spitting it out.
   b. Using the stirring spoon, take a small amount (no more than 1/2 teaspoon) of the sucrose mixture and dump it onto another spoon. Taste from this spoon so you don't contaminate the rest. Hold the mixture in your mouth, tasting it with your tongue, then spit it out into a cup or the sink
   c. Record your observations regarding any flavor you detect.

7. Repeat Step 6 for each solution, making sure to rinse your mouth between each tasting.

8. Now rate the sweetness level of each. Give a rating of one to the solution you thought tasted most sweet and a rating of three to the solution that was least sweet.

Spit Cup: While it sounds gross, and might even BE gross, rinsing your mouth helps prevent your body from becoming fatigued and changing your perception. Your taste buds sense differently once food enters your stomach. Rinsing also helps to minimize any carryover sweetness or flavor from the previous tasting.

9. Summarize your results in the table.

Sweetener Source	Solubility	Flavor	Sweetness	Rating
Sucrose				
Fructose				
Corn Syrup				

10. Ask one or two other people to also rate the sweetness levels—without talking to one another or looking at each other's ratings. It's easy to be biased by someone else's opinion if you talk about what you're experienc-

ing, which is why in sensory analysis, we make our ratings individually, without talking to other people.

11. Share your observations with one another. Did you all rate them the same?

Carbohydrates Experiment #1 Discussion: Sucrose vs. Fructose Experiment

Sucrose and fructose are very different molecules. Fructose is more *soluble* in water than sucrose, which you likely noticed as soon as water was added to the crystals. Even the crystals themselves act differently. Fructose will absorb moisture from the air readily in a humid environment. The increased solubility of a sweetener usually results in an increased perception of sweetness, as is the case for fructose. When you tasted the solutions of both fructose and sucrose, you likely noticed this phenomenon, but not all people have the same ability to taste. Don't be concerned if they tasted similar to you!

Soluble: The ability or readiness of a solid to dissolve in a liquid. For example, another way of saying that salt dissolves in water is to say that salt is *soluble* in water.

Corn syrup is a blend of fructose, glucose, and water. As we just discussed, fructose is more soluble than sucrose, resulting in a stronger taste of sweetness. Glucose is less soluble than sucrose. It's generally perceived to be less sweet than either fructose or sucrose. Depending on the ratio of fructose to glucose in your sample of corn syrup, you may have rated corn syrup either a two or three.

You also likely noted some differences in how people rate the sweetness levels. Our abilities to taste are not the same, which should not concern you now, but as you may imagine, this makes a food scientist's job more challenging. Disparity in perception among people is a real issue and must be taken into account.

Consumer sensory testing is an important part of developing food products. This exercise was a very simple example of sensory testing. One possible career choice in food science is focusing on making sure that consumer testing is set up appropriately to produce valid results for product developers so that they can appropriately formulate their products.

Carbohydrates Experiment #2: Cornstarch vs. Potato Starch

Background

Starches are quite different depending on their source.

Items Needed

- 4 tablespoons cornstarch
- 1 medium-sized white potato, peeled and cut into approximate 1/2-inch cubes (different potato varieties will produce somewhat different results, but any will do for this experiment)
- Water
- Small saucepan
- 4 plates (any type will work except paper)
- 2 freezer-safe containers with lids
- Marker or pen
- Masking tape for labeling containers

Note:

You'll need to complete steps 1 through 3 prior to recording observations, unless you're working with someone else who can take notes.

Procedures

- Cornstarch

1. Label the plates "Cornstarch, Gelatinized," "Cornstarch, Gelatinized, Broken Granules," "Potato, Gelatinized," and "Potato, Gelatinized, Broken Granules."

2. Label the containers with lids: one "Cornstarch, Gelatinized, Frozen" and the other "Potato, Gelatinized, Frozen."
3. Add 1 1/2 cups of cold water to the saucepan and stir in 4 tablespoons cornstarch with a spoon. Let the cornstarch settle, then stir it again. Record your observations.

> ***Note:*** Be careful during this step because the starch will tend to pop as it comes to a boil, sometimes ejecting hot globs of starch that can burn your skin!

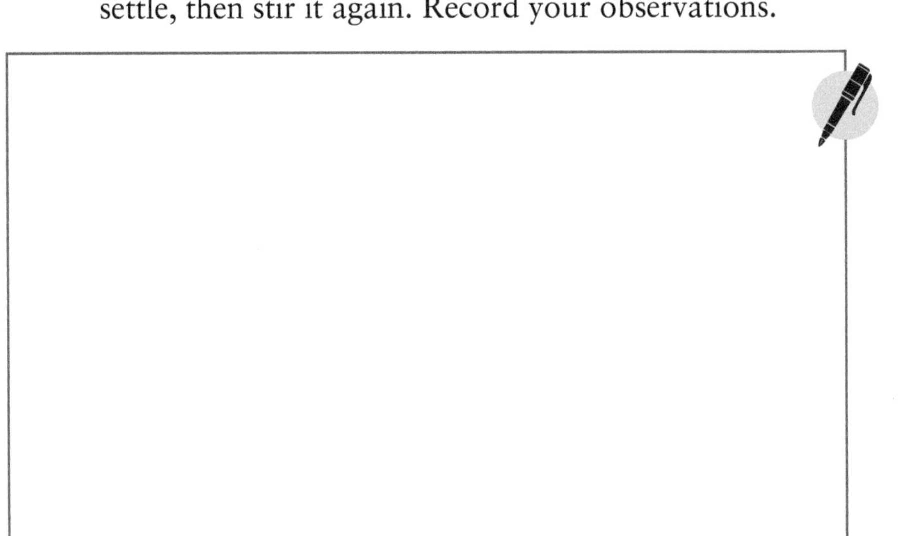

4. Bring to a boil slowly on medium heat, stirring slowly but constantly. The starch will begin to thicken. Continue to gently stir until a full rolling boil is achieved. You should continue to stir constantly but gently, once a full boil is reached, for 2 minutes. Note the consistency. Turn off the burner and remove the pot from the heat. What food product does this look like? Record your observations.

5. Carefully spread a 1/8inch thick or thinner layer of product onto the plated labeled "Cornstarch, Gelatinized."

6. Add additional product to the plastic container with a lid labeled "Cornstarch, Gelatinized, Frozen." Place the container with the lid attached into the freezer.
7. Use a hand mixer to beat the remaining product. Were there any changes noted? Can you explain what's happening?

8. Spread this product into a 1/8inch thick layer onto the plate labeled "Cornstarch, Broken Granules."

- Potatoes

1. Place the 1/2-inch potato cubes into an excess of water so that at least 1 inch of water is below the floating potatoes. Bring to a boil, reducing the heat and simmering until the potatoes are very soft and start to come apart in the water.

2. Pour off and discard the starchy water. Mash the potatoes carefully and minimally, just enough to destroy the cube shapes. What's the consistency? Divide into two equal portions, then spread about 1/2 of one portion into a 1/8inch thick layer onto the plate labeled "Potato, Gelatinized."

3. Add the rest of the first portion to the second freezer container, labeled "Potato, Gelatinized, Frozen." Place the container with the lid in the freezer.
4. Beat the remaining potatoes with a mixer on medium to high speed for 2 minutes. Can you explain what's happening? What's the consistency? Record your observations.

5. Take this product and spread it into a 1/8inch thick layer onto the plate labeled "Potato, Broken Granules."
6. For both the corn starch and potato starch plates, let sit and air dry, observing them over several days. What happens to the consistency? Can you explain what's happening? Record your observations.

7. For both the cornstarch and potato containers with lids in the freezer, remove them after at least 5 days in the freezer and let them thaw at room temperature.
8. Evaluate the texture of both. Press on the thawed starch with a large spoon. What happens? Record your observations.

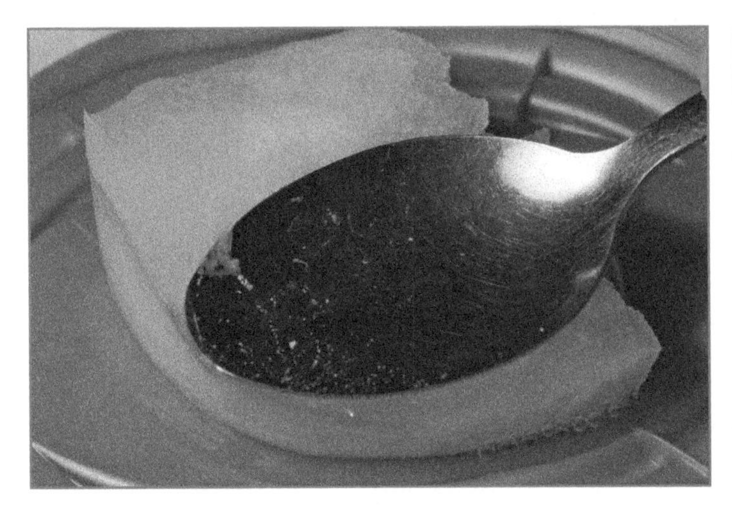

Carbohydrates Experiment #2 Discussion: Cornstarch vs. Potato Starch

Cornstarch has a higher amylose:amylopectin ratio than does potato starch, so fully hydrated potato and cornstarch have quite different textures. When fully hydrated and *abused*—mechanically agitated until the granules begin to burst—the differences become even more dramatic.

In your kitchen: Perhaps you've transported a granola bar in your backpack, only to discover that it didn't survive the trip well. Your transport system wasn't sufficiently robust to prevent abuse of your granola bar. At least now it might make a good topping for yogurt!

Prior to starch granule breakage, the viscosity increase is largely from friction between the swollen granules. After breakage, there can actually be a reduction in viscosity as the granules are destroyed and individual hydrated starch molecules become freer to move around and stretch out. The dispersion's consistency changes its character. In the case of potato starch, the higher amylopectin content results in a stringy, unpleasant, and almost mucous-like texture when the gelatinized starch is abused, as we did with the mixer. You definitely shouldn't do this to mashed potatoes at your holiday meal!

After drying, you'll notice that the corn and potato starch end up with very different textures. For the fully hydrated but not broken starch, thin, clear, and shiny films may develop that are brittle in texture. These films may look and feel like plastic wrap, but eventually become very brittle and will crack. You may have seen this after cooking a starchy grain, such as rice. A sticky film will remain in the pot, but if left to dry, it becomes like a brittle piece of plastic and easily flakes off.

It's interesting to note that as plastic is made of polymers, so is starch. It's a similar phenomenon that is occurring. Edible starch films such as these are even used in food products for various purposes. For potato starch, films may not even develop, but the starch will seem to pull together as it dries, which is additional evidence of the radically different starch characteristics between corn and potato starch.

When you freeze and thaw hydrated starch, the action of freezing causes some swollen starch granules to burst due to

Ingredient abuse: Over-stirring gelatinized starch can cause unwanted changes in texture. Ingredients can be abused in many ways, and this is an important concept in food processing. Another example is transportation of ingredients from one point to another. Many types of transportation can cause breakage. Depending on the product, this damage may not make a difference, or it may make it impossible to create a quality product.

the formation of ice crystals. You'll notice that in both the corn and potato starches, a small amount of water separates from the hydrated starch. Also, the hydrated starch itself will become tough, especially with the higher-amylose-content cornstarch. Hydrated starch toughening and contraction is due to a phenomenon called *retrogradation*, where the amylose molecules reorient themselves and come together again, but associated in different ways. For both of these reasons, the starch matrix will lose its ability to hold water. When you press on it with a spoon, water will flow out, almost like a sponge.

Notice that when the potato starch is pressed with the spoon, not as much water came out. In addition, the thawed mashed potato mixture will likely have a similar texture to when it was fresh. These observations are due to the differences between potato and cornstarch.

Retrogradation is also one of the same phenomena that helps describe the process of bread becoming stale. More will be discussed on this in the *Carbohydrates* workbook. If you're particularly interested, after thawing, you can put the starch pastes back in the freezer, remove them, make your observations, and repeat. Do this three to four times, with several days in the freezer each time, and you'll note that the effects described above become even more dramatic. This freeze/thaw cycle, as noted before, is a significant problem for the development of frozen food formulas. Starches can be modified to prevent these defects associated with drying or freezing.

This experiment was very simplified. While you can eat these starch pastes or gels—if you were careful to prepare them in a sanitary fashion—they won't be very appetizing. In the example of cornstarch, it is essentially what you buy as pudding in a box from the grocery store. However, to make it a dessert, you'd also need a sweetener and flavor of some kind. Sweeteners, flavorings, or anything else affect the ability of starch granules to hydrate, and affect how they'll behave after hydration and over time. Food systems are incredibly complex and require a significant amount of work to create something with a pleasing texture and flavor that will last long enough for a consumer to purchase it and ultimately consume it.

WHAT DO YOU THINK?

Look at the side panel of packaged foods in your kitchen. Do you see any that contain *modified starch*? This will usually be cornstarch. What role do you think the modified cornstarch plays in the food? Call the toll-free number listed on the package and ask the manufacturer what type of modified starch is in the food, and what role it plays. You may or may not get an answer right away. However, you'll eventually get some type of an answer if you indicate that you really would like to know! For example, if the person doesn't know, ask them to contact you when they find someone who does know. Sometimes companies won't share exactly what's in their formula because it's the secret to their success, and they may think you're a competitor trying to copy them!

JOURNALING IDEA

Now that you know a little about carbohydrates, write about your favorite one carbohydrate molecule. Why is it your favorite? Describe what you can about its structure and what you know about that structure that may determine the characteristics you enjoy. Write about something you can make in the kitchen that may have a similar structure, and therefore that you may also like.

CHAPTER REVIEW

Write short essay responses:

1. Why does a dry cornstarch film look and feel so much like plastic wrap?
2. Describe a starch granule's thoughts as if it were alive while being boiled, gelatinized, and fully hydrated, and ultimately becoming part of butterscotch pudding. This was the starch granule's life goal.
3. Write a short story about your favorite mostly carbohydrate food, including why it's your favorite. How much fat or protein does it contain that help it to be your favorite carbohydrate?
4. Describe one area of your life that doesn't involve chemistry in some way.

CHAPTER 7
PROTEIN

If carbohydrates are the staff of life, proteins are what build the staff. In biology we learn that proteins are involved in every aspect of life, all the way down to what goes on inside the cell. Proteins are instrumental in the production of energy within a cell, converting what we eat into a form of energy that the body can use. Proteins maintain and replenish the body, including replenishing themselves. Very simplistically, enzymes, which are proteins, run our bodies, ultimately under the direction of our DNA. With that in mind, we'd better consume some protein to keep all that going!

Let's examine some interesting aspects of protein and have some more fun with experiments. You'll see that in addition to being pivotal to life, proteins are also very important to the flavors and textures of most foods we love.

STRUCTURE

Proteins are made up of about 20 amino acids. Each amino acid has the same general structure as shown at right, with the "R" representing a side chain that's different for each amino acid. This *R group* is what determines each amino acid's functionality, including how it reacts with neighboring amino acids and other molecules.

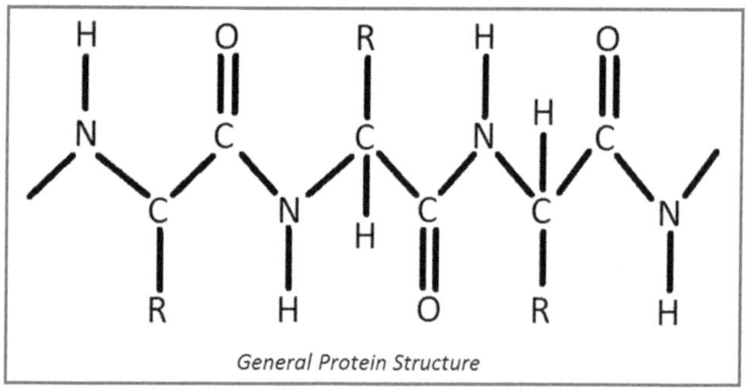

General Protein Structure

Protein general structure[61]

Amino acids are joined together in many different sequences. The arrangement of amino acids in a chain comprise the primary structure of a protein. Secondary and tertiary structures indicate how an individual chain folds and interacts with itself.

Quaternary structure defines how more than one protein chain will interact with each other, forming an even more complex structure. Primary, secondary, tertiary, and quaternary levels of structure together make up a three-dimensional protein with unique biological and chemical properties. The complexity of protein functionality is one of the most interesting aspects of nature!

> ***Protein Levels of Structure***
> ***Primary (first):*** Combination and sequence of amino acids
> Secondary (second): Local three-dimensional structures within a single protein
> Tertiary (third): The three-dimensional shape of a protein in its natural state.
> Quaternary: The three-dimensional shape of two or more proteins interacting with one another to form a functional group.

DENATURATION

A disruption of any structural level will change the protein's functionality. Together, these types of disruption are commonly known as *denaturation*, and the word describes the usually irreversible loss of any of the secondary through quaternary structures. Usually, denatured protein doesn't act the same as a protein in its native state. This can occur in many ways.

For example, protein structure is very sensitive to changes in pH (the level of acidity), temperature, mechanical action (such as beating with a mixer), freezing, and ionic concentration. Denaturation isn't usually reversible—or at least not completely reversible. A great example of protein denaturation is egg whites. The consistency and appearance of raw vs. cooked egg whites is dramatic. Just try to make a cooked egg white revert to the raw stage—it doesn't work! Other examples of protein denaturation in food are cheesemaking and the change in texture of raw beef when it's cooked.

ENZYMES

Denaturation of proteins becomes important in understanding food processes. Enzymes are a great example again. Enzymes bring together two more components of a chemical reaction and facilitate that reaction. In the case of enzyme catalyzed reactions, the components the enzyme brings together are called *substrates*. Below is a simple illustration:

Enzyme (E) + Substrate (S) together form a complex, which results in the Enzyme and a Product (P):

$$E + S \rightarrow ES = E + P$$

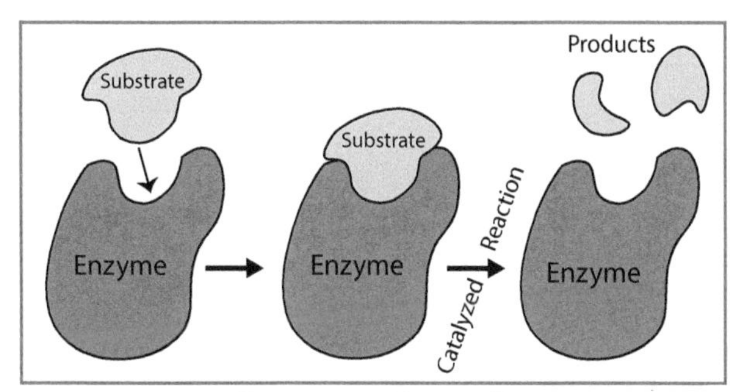

Enzyme + substrate[62]

Enzymes are present everywhere in nature and exist in both plants and animals. Sometimes it's desired that naturally occurring enzymes be allowed to continue their work after harvesting, while at other times it's desired to completely stop any enzymatic activity. For example, enzymes contribute to continued ripening of harvested fruits, which is desirable up to a point...then you want it to stop.

Video Box: An interesting video discussing enzymes and denaturation: https://www.youtube.com/watch?v=Z_ZieKjEQ7s

PROTEIN INTERACTION

Rising bread: Why does bread continue to rise in the oven if all the yeast has been killed? There are at least a couple of reasons:

1. It takes a while for the yeast to be killed, so it'll continue to act for a short time, even while in the oven.
2. The pockets of carbon dioxide within the dough are just pockets of gas. All gases expand as they heat up. This also causes dough expansion—at least while the dough is still elastic.

Another great example protein's role is traditional wheat flour–based bread. The basic ingredients of bread are wheat flour, yeast, sugar of some type, and sometimes salt. Yeast is a living organism that metabolizes (eats and digests) sugar. A byproduct of that action is carbon dioxide gas, which is what makes bread dough rise when it's kept at warm temperatures for a while. At high temperatures, such as when baking in an oven, the yeast is killed and the bread structure sets. The structure that allows the loaf of bread to retain its shape and not collapse is protein-based and is referred to as *gluten.*

Glutenin + Gliadin + Water + Mechanical Action = Gluten

Gluten is a combination of two classes of proteins found in wheat: glutenins and gliadins. These two classes of proteins combine to form gluten when they interact. Kneading bread dough by some mechanical action, either by hand or machine, forces the glutenins and gliadins to interact with each other. If there's insufficient mechanical action, not enough gluten will develop to hold the structure once the loaf of bread is removed from the oven. In this situation, the bread may collapse!

Bread dough[63]

ANIMAL PROTEINS

Plant-based proteins are one thing; animal proteins are another entirely. The primary source of animal protein used in human foods are the muscles of animals. However, food products have been developed from almost every other part of an animal that contains protein, including the skin. Muscles are what move the skeletal structure of an animal. They attach to bones via tendons and are associated with other *connective tissues* that are between sections of muscle, among other things. Ultimately, connective tissues form the harness that *connect* muscles to tendons and then to the bone.

Angus beef[64]

Whole meat can be processed in many different ways, resulting in entirely different finished foods. It's truly amazing.

Consider purchased ready-to-eat whole meats. They can be cured, smoked, fermented, dried (such as *beef jerky*), or simply cooked and sliced. Other meat products, such as hot dogs, Spam® (Hormel Foods), Vienna sausages, and bologna are entirely different and unique in their own chemistry compared to whole-meat products, even when they both may start, for example, from beef.

Hundreds and thousands of meat products are available around the world, from fish to fowl and everything in between.

EXPERIMENTS: LET'S MAKE A MESS!

Protein Experiment #1: Enzymes and Apple Browning

Background

In this experiment we'll explore a reaction involving proteins that will be familiar to most everyone—that of surface browning in fresh fruits and some vegetables. We'll also explore ways in which this browning can be controlled, and learn a bit more about proteins in the process.

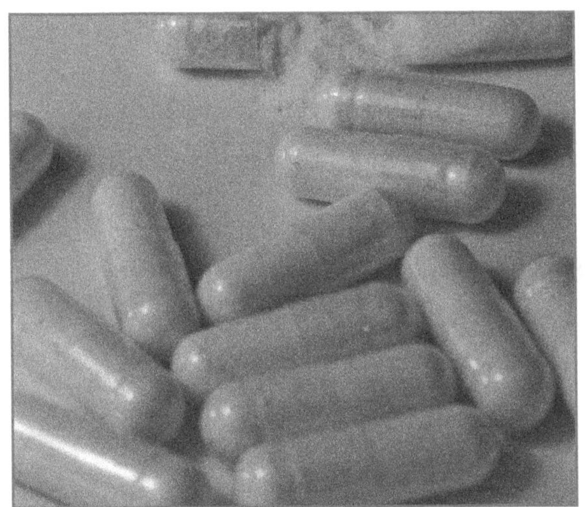

Items Needed

- 1 medium-sized apple, room temperature—NOT cold
- 1 small pear, ripe
- Water
- Lemon juice
- 500mg calcium ascorbate capsules (these can be purchased at grocery stores, or online)
- 1 large dinner plate
- 3 medium-sized bowls
- Masking or painter's tape for labels
- Marker or pen

Procedures

The browning reaction will begin immediately. The purpose of this experiment is to demonstrate differences in cut fruit sections' treatment, so you want to treat them immediately. Please read through the procedures and prepare everything that's needed before cutting.

1. Label the plate in quadrants as follows: "Non-Treated," "Water Rinsed," "Lemon Juice," and "Calcium Ascorbate Dipped."
2. Label the bowls as follows: "Lemon Juice," "Calcium Ascorbate," and "Water"

3. Prepare the lemon juice solution by adding 1 1/2 cups of water to the bowl and adding 1/3 cup lemon juice.
4. Prepare the calcium ascorbate dip by adding 12 capsules to 2 cups water in the bowl. This can be done by pulling the two capsule halves apart and dumping out the contents. Stir vigorously with a whisk and let sit. Skim off and discard any surface material that didn't dissolve.
5. Cut the apple into about 3/4-inch pieces, with the core removed.
6. Immediately put about 1/4 of the apple pieces in each of the two bowls, stir gently, and let sit for five minutes. While these are sitting in their bowls, rinse another 1/4 of the apples thoroughly under running water. Along with the last 1/4 apples that weren't treated at all, re-

move and place all the treated apples onto their corresponding plate sections.

7. Repeat the last step for the pear, leaving it in slices if desired.
8. Start a timer or note the time. Record your observations of the fruit and their appearance at the following time increments: 0 minutes, 30 minutes, 2 hours, 4 hours, and 6 hours.

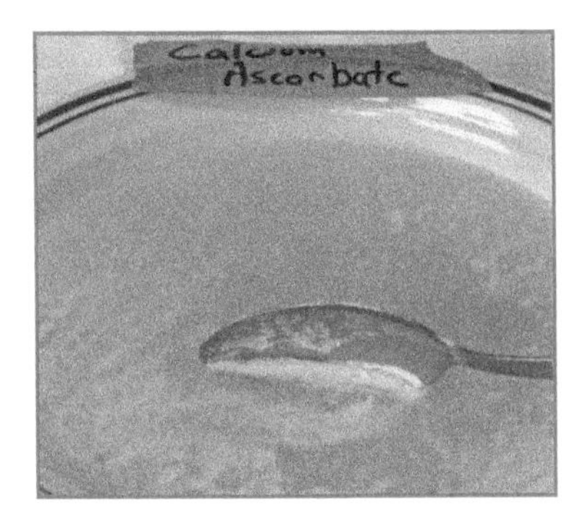

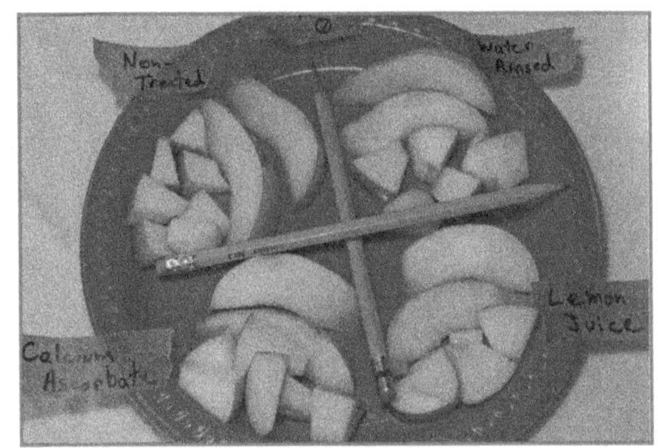

Time Zero

2 hours

06 hours

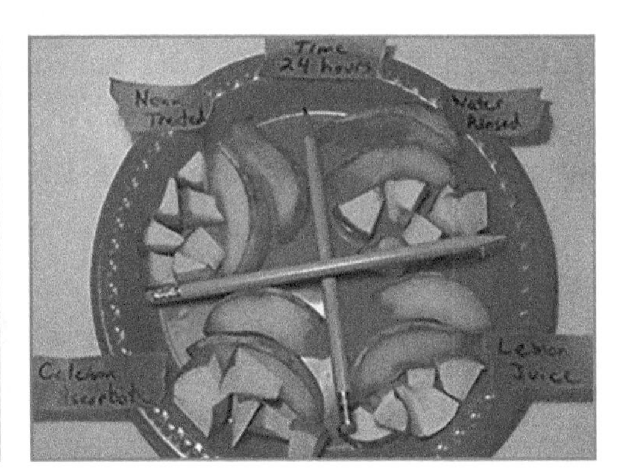

24 hours

You may want to take a picture of the experiment as you start to help you remember what they looked like at "time zero." You may even want to do this at other points throughout the experiment.

Observations:	
0 Minutes	
30 Minutes	
2 Hours	
4 Hours	
6 Hours	
24 Hours	

Protein Experiment #1 Discussion: Enzymes and Apple Browning

All enzymes are made of protein and are extremely important for life processes and for making food. This experiment demonstrates the activity of a common class of enzyme, polyphenoloxidase. While there are multiple variations of this enzyme, for ease of description we'll refer to it as a single enzyme. Polyphenoloxidase is responsible for many reactions that we see on a daily basis, including the browning of apples and other fruits. However, it's possible to control and slow the enzyme reaction to some extent. As discussed previously, enzymes have a three-dimensional structure. They also have operating conditions at which they function best. Understanding what these conditions are, and sometimes even the best structure, makes it possible to disrupt the enzyme's action.

Polyphenoloxidase is naturally present in the apple's cells. When it has access to oxygen, it speeds up a reaction with proteins in the apple that ultimately generates a brown color. When an apple is cut, the apple flesh's cells are disrupted. The polyphenoloxidase present on the surface of the cut apple now has access to oxygen in the air and other substrates that are needed to generate the brown-colored compounds. These browning reactions are catalyzed by the enzyme and begin immediately.

However, it's possible to slow down the reaction by washing away some of the enzyme and other substrates that are present on the cut apple's surface. It's even more effective to wash the surface with an acidic solution, which moves the system out of the optimal operating range for polyphenoloxidase. Calcium ascorbate also disrupts this activity, and temperature is another way to control reaction time. Generally, warmer temperatures produce faster reactions, which is why it was emphasized for this experiment to have a room temperature apple instead of a cold one.

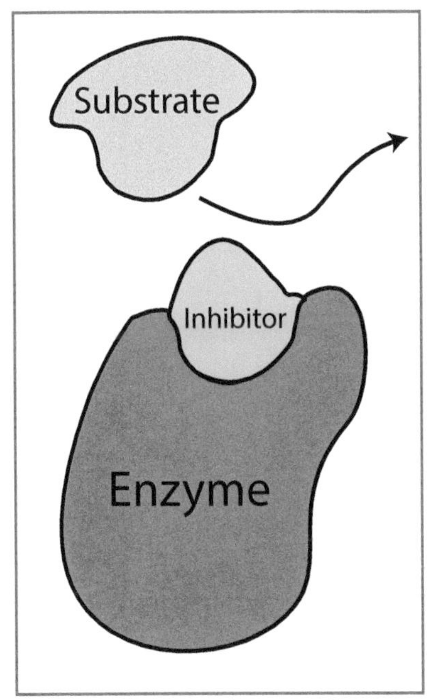

Enzyme + disrupter[65]

If you were careful in processing the apples quickly, you should see a marked difference between the apples based on their treatment. The apples that weren't treated at all should have begun browning very quickly, within minutes. Those rinsed in water should have browned more slowly. Why? We rinsed some of the enzyme polyphenoloxidase away, along with some other substrates needed to accomplish this enzymatic browning. Apple slices soaked in water likely browned more slowly than those just rinsed in water. The apples soaked in calcium ascorbate should have browned the most slowly due to its interference with enzyme activity.

Interesting, isn't it? Can you think of other things that might affect the rate at which the enzyme can react? For instance, have you ever purchased apple slices from McDonalds®? Their apple slices are always bright white, not having hardly browned at all since they've been treated with calcium ascorbate. Calcium is a necessary mineral for human good nutrition and ascorbate is related to ascorbic acid, which is vitamin C, another vitamin important to the body. Consequently, calcium ascorbate is a great, natural solution to inhibiting browning and maintaining the fruit's appearance for a longer period of time.

While the above discussion revolved around an apple, you likely noticed a disparity between the relative browning of the apple and the pear. The flesh of both of these fruits differ greatly, including in the amounts of protein, rates of oxygen permeability, and the type and quantity of sugars they contain. Dissimilarities such as these are partly what make food science, and in particular food chemistry, complex and exciting!

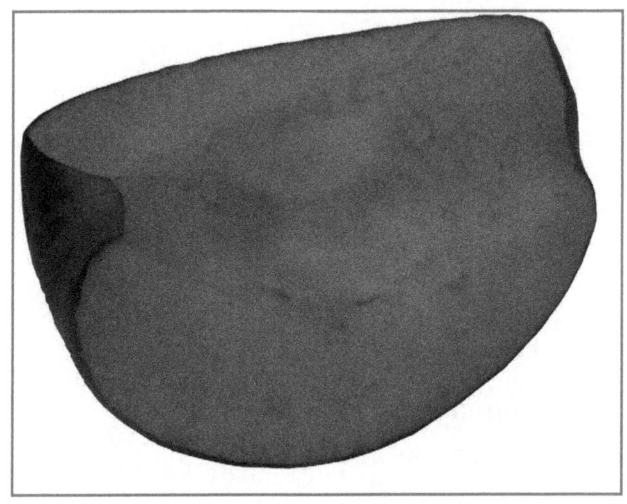

Browned apple[66]

Protein Experiment #2: Onions Can Make You Cry…or Not!

Background

In this experiment we are again going to examine enzymes. We'll learn about how enzymes in an onion react with other components and produce a compound that irritates the eye, and how we can minimize this effect.

Enzymatic Browning:
https://www.youtube.com/watch?v=Tt9FYHmMOjU

Items Needed

- 1 small onion
- Sink with running water
- Small sharp knife
- Small cutting board
- Large baking dish

Procedures

1. Fill the large baking dish with cold water. Insert the wooden cutting board and submerge the small onion. Carefully cut it in half while it's underwater. Leave both halves in the baking dish.

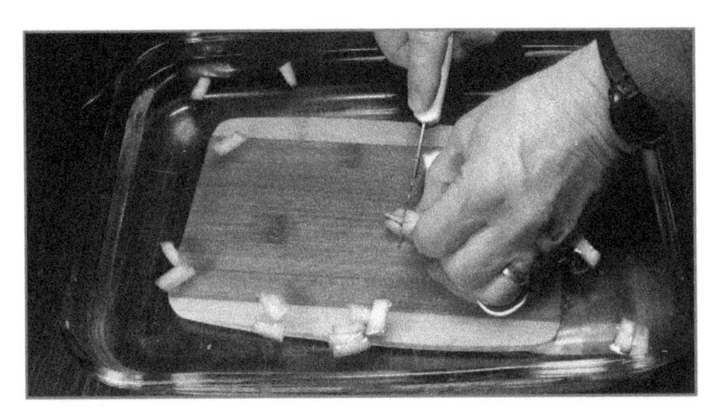

2. Cut 1/2 of the onion into smaller pieces while it's still submerged in water. Note any physical response your eyes may be having. Can you smell the onion? Record your observations.

3. Remove the cutting board from the water, rinse, and dry.
4. Remove the other half of onion from the baking dish of cold water and cut it on the cutting board into small pieces. Note any physical response that your eyes may be having. Can you smell the onion more or less than before? Do you notice a difference in how your eyes feel or a difference in the smell? As with most everything, humans differ widely in their response to these compounds.

…cussion: Onions

…ontain many different types of
…erform various functions. One
…tates a reaction that converts
…rally present in onions, into
… through the air, it eventu-
…ch become very irritated,
… do their job and try to

… the same constraints
…perating conditions
…y for the reaction.
… for an enzymatic
…ng the substrate
…prevent it. This
… a stream of water.
… crying!

Protein Experiment #3: Some Cooking Methods and Their Effects on Beef

Background

Animal proteins are complex and interesting. Individual muscle fibers are formed into bundles, small bundles are formed into larger bundles, and these bundles are all ultimately held together by connective tissue that forms a whole muscle. How you cook this muscle has a great effect on the finished texture. In this experiment, we'll explore three methods for cooking the same piece of meat and evaluate the finished product. Any cut of meat will do. In this case, I used sirloin steak.

Items Needed

- One 8 oz. or larger cut of beef, at least 1 inch in thickness
- Frying pan
- Saucepan with lid
- Small oven dish with lid (aluminum foil can also be used as a lid)
- Knife
- Plate
- Cutting board
- Spatula
- Water
- Labels
- Pen or marker

Procedures

Important for this experiment is to cook the meat so that they all can be evaluated at approximately the same time. Read through all the instructions and determine when you need to start each experiment to make that happen.

1. Place the meat onto the cutting board and cut it into three equal pieces.
2. Preheat the oven to 225°F.
3. Place one piece into the oven dish, cover it with the lid or foil (if you use foil, make sure to crimp it to the container sides to minimize moisture loss), carefully place in the oven, and roast for approximately 2 1/2 hours, or to an internal temperature of 150°F.

4. After 2 hours have elapsed while roasting the first piece of meat, begin preparing the other two pieces of meat.
5. Place the second piece of meat in the saucepan and cover it with 2 inches of water.
6. Cover with the saucepan lid and bring to a boil on medium heat. Reduce heat and continue to simmer for 10 minutes. Turn off the heat.

7. Heat a frying pan on medium heat. You may use a little bit of oil in the pan if desired. Place the meat in the pan and *sear* it on both sides until the meat center is only slightly pink.

Sear: To cook on a high temperature surface for a short period of time. In this case, the result is a browned/blackened surface with an interior that can be raw if seared for a very short period of time.

8. You can flip the meat over several times to cook it evenly. Turn off the heat, remove the meat from the pan, and place it on the plate.

9. Remove the meat from the saucepan and water and place it on the plate.
10. Remove the oven dish from the oven, remove the meat, and place it on the plate.

11. Let all the pieces of meat cool sufficiently for tasting, then make your evaluations of all three.

Method	Texture	Flavor	Preference
(rank 1, 2, 3)			
Frying Pan			
Boiled			
Oven			

Protein Experiment #3 Results and Discussion: Some Cooking Methods and Their Effects on Beef

You may have heard of pot roast, which is what we created with the first cut of meat in the oven.

The connective tissue within muscle is broken down at different rates depending on time, heat, and moisture. Frying is considered a dry method, while boiling and roasting in a covered pot are considered moist methods. Collagen is one of the primary connective tissues found in meat.

Collagen: An insoluble fibrous protein that is the chief component of connective tissue, including skin, tendons, and even bones. When heated with water for long periods of time, collagen denatures, resulting in gelatin, and also can be used to make glue.[xv]

With sufficient time and availability of moisture, collagen is hydrolyzed into gelatin. Frying completes the cooking too fast to do this. Consequently, the collagen is primarily present in its native form. In this cut of meat, it would have resulted in a fairly tough texture. Most people find the flavor using this method to be pleasant, but they sacrifice tenderness.

xv "Collagen." Merriam-Webster.com. https://www.merriam-webster.com/dictionary/collagen. Copyright 2019 Merriam-Webster, Incorporated. Accessed 04 February 2019.

Boiling provides moisture and heat. However, the length of time in this experiment was inadequate to break down the collagen. The product would still have been relatively tough. Most people also find the flavor of boiled beef not as pleasant as the other two methods of preparation.

Pot roasting is another moist method. Although it may seem like a dry method, with the dish covered, sufficient moisture is available to consider it to be moist. Roasting at a low temperature for a long period results in sufficient time to break down the collagen and results in very tender meat that falls apart into individual muscle strands. The meat will generally be very moist. Most people find the flavor of this method of preparation pleasant and enjoy the texture. Slow cookers are also a great way to prepare a roast in this fashion.

WHAT DO YOU THINK?

What about the color changes in the meat? It was red and became gray or brown depending on the preparation method. What's going on there? Have you ever reheated cooked meat? It doesn't taste anywhere close to freshly prepared meat. What about meats that can be stored in a bag at the checkout line? Why can they sit like that, without refrigeration? Why are bologna and some hot dogs pink?

Hot dog and pepperoni[67]

What other enzymes are used in food production? What role do they play? Do we have to eat animal protein in order to survive?

As you can tell, we've just scratched the surface of proteins in food science. This chapter was only intended to give you a taste (pun intended) and perhaps stimulate some curiosity and a desire to find out more!

JOURNALING IDEA

How many different ways can an egg prepared? Ask a cook you know or do a web search regarding the question and write down what you find. How is it possible that the same egg, when prepared differently, can have textures such is found in boiled egg whites or scrambled eggs, or even a hard or soft foam such as meringue on a pie? How can your new knowledge of protein help explain this?

CHAPTER REVIEW

Write short essay responses:

1. What is your favorite protein source and why? How do you like that protein prepared? Describe what you can regarding the science behind what happens to the protein as it's prepared and finally becomes the food you enjoy.
2. American citizens eat a lot of beef, which comes from cattle. Cattle eat only plants. How do cattle convert plants into protein? Can humans do this?
3. An effective way to sterilize something is to expose it to a radioactive source, which will effectively kill any living process. Spices are often treated this way to reduce microbial activity, since heating them would destroy, or at least change, their flavor. From your understanding of proteins, can you think of why radiation is so effective? Hint: Radiation of the proper type denatures proteins.
4. Describe what you think the scientific mechanism may be for the different results seen in preparing beef from Protein Experiment #3. Can you think of other proteins where a similar mechanism might be employed to achieve a desired result?
5. Think of an animal muscle. Tendons, muscle fibers, and other connective tissues are all mostly protein, but they look and act quite differently. How can this be when they are all made of protein?

Field Trip: Visit your local grocery store and look in the baked goods aisle to find products labeled Gluten Free. Celiac Disease afflicts many people. People with this condition are unable to digest gluten. If these individuals eat products with gluten, they experience intestinal distress, among other things. Other people have health issues associated with gluten, or wheat in general, although they aren't as severe. Food companies have responded by developing foods without gluten. This can be very challenging for a food scientist.

CHAPTER 8
FAT (LIPIDS)

Butter[68]

***Fat.* In today's lexicon, even** the word sounds bad! However, did you know that fat is necessary for life to exist? Without fat, you'd most certainly die.

From a food science standpoint, fat is a very interesting and versatile material with which to work. It can be made into a finished product ready for sale, such as olive oil. It can also be incorporated into many other products as an ingredient, such as in baked goods like muffins. Fat can also be used as a medium in which to prepare other foods, such as frying and producing potato chips. Here we'll briefly explore the world of fat and food, which should get you thinking about the foods that you see and eat.

WHAT IS FAT?

Glycerol and Fatty Acids

Fat is a common name for the scientific term *lipids*. Lipids are compounds that are generally not soluble in water, but are soluble in organic solvents, such as ether.

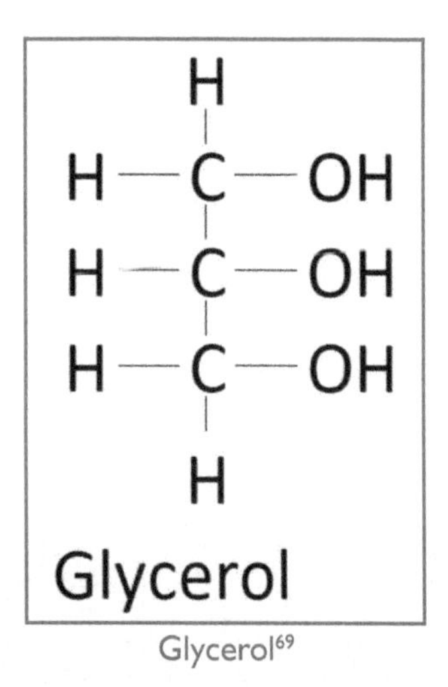

Glycerol[69]

Glycerol is the name given to the pictured molecule. It's simple, yet so extremely important in that it's the base for all fats. Glycerol is also referred to as *glycerin.* Take this same glycerol backbone used to make fat and replace the hydrogens with nitrogen- and oxygen-containing compounds, and you get nitroglycerine, which is explosive. *Boom!*

Fats are composed of fatty acid chains on a glycerol molecule backbone. Each glycerol molecule can hold up to three fatty acid chains.

When there's 1 (mono) fatty acid chain on a molecule, it's known as a *monoglyceride.* When there's 2 (di) of the three spots on the molecule, it makes a *diglyceride*, and 3 (tri) define a *triglyceride.* You may have heard those terms before because they're common in discussions regarding fats and diet, and overall health.

Fatty acid chains themselves consist of carbon chains of different lengths, and the chemical bonding within the chain can be different. For example, some chains contain carbon atoms that are single-bonded to each other, and others are double-bonded. Those carbon chains that are single-bonded are straight. Those that have double bonds will have a "kink," or a bend, in the molecule. You may have heard the terms *saturated fat, unsaturated fat*, or even *polyunsaturated fat.* When a fatty acid is fully hydrogenated, it means that all the carbon atoms have a hydrogen atom singly bonded at each possible site. In other words, it's *saturated* with hydrogen atoms and that situation results in a straight fatty acid.

Fatty Acids: A particular type of acid called *carboxylic acid* followed by a chain of linked carbons of varying lengths and characteristics, which together make up the molecule's functional properties.

When you remove 2 hydrogen atoms, you end up with 2 carbon atoms double-bonded together and the fatty acid is no longer saturated.

Polyunsaturated fats contain more than one location of double-bonded carbon atoms, resulting in multiple "kinks" in the chain.

Triglyceride with Two Saturated and One Monounsaturated Fatty Acid

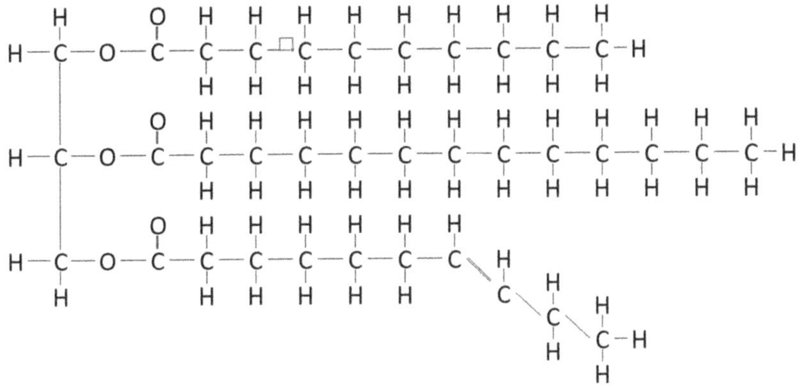

Triglyceride structure2[70]

The number of chains on a molecule, and the length and characteristics of those chains—including the incidence of double and single bonds—are the cause of physical differences we see from various fat sources. These differences are what result in the functions, or functionalities, of diverse fat types. An understanding of this underlying science can help the food scientist formulate food products with desirable characteristics.

Trans Fats: This description of fats has to do with the different shapes that two chemically identical molecules can take. When fatty acids are artificially hydrogenated, if care isn't taken to fully hydrogenate them, sometimes they can result in a fatty acid in the *trans* configuration. In this case, even though a double bond exists, the fatty acid will still tend to be straight. Straight molecules, whether saturated or trans-unsaturated, have been identified in current health science understanding as contributing to heart disease.

FAT'S ROLE IN FOOD: FUNCTIONALITY

A good analogy for fats is stacking lumber. The long, straight fatty acid chains can be thought of as straight pieces of lumber, and unsaturated fats as bent or curved lumber. It's much easier to make a neat and orderly stack of lumber out of straight lumber than curved lumber. When the fat chains are all laid out straight, the fat they create is a solid. But when the fat consists of bent and curved chains, it's a liquid.

Vegetable oils generally have a large degree of unsaturation, with many bends or kinks being present. This means it's hard to make it solid because the fatty acid chains don't line up together. They get in the way of one another.

It takes deliberate effort to make this type of fat solidify. In the case of some unsaturated (bent) oils, a refrigeration compressor removes heat from the oil, which allows the fat molecules to solidify. Saturated fats will solidify at higher temperatures.

Sometimes fats will be liquid even at refrigeration temperatures. Some oils must be below the freezing temperature of water to become solid.

Shortening contains a blend of oils that's solid at room temperature. Some of these blends may contain a *fully hydrogenated* oil. Hydrogenation means that the kinks or bends in the fatty acid chains have been removed by forcing hydrogen into those bonds. This chain straightening allows for fatty acid chains to line up together much more easily, resulting in solidity at higher temperatures.

Liquid and solid fats act very differently in foods and they leave a very different feel when consumed. There actually is an industry term for what foods feel like in your mouth! Luckily, it isn't hard to remember and it's intuitive: *mouthfeel*. The mouthfeel of liquid vs. solid fats is very different, depending on the food temperature, and it's also important to consider the temperature of the mouth it's in. For example, some less expensive chocolates combine cocoa bean fat with another fat, resulting in a higher melting point.

Cocoa bean fat generally melts just below body temperature, which gives it a characteristic mouthfeel. The substitute fats of the cheaper chocolate, with a higher melting point, will have a waxy mouthfeel that many find unpleasant.

Shortening: What's in a name? How did shortening get its name? Earlier gluten formation was introduced that required mechanical mixing or kneading of wheat-based doughs.
When fat is introduced, it can prevent some gluten formation by not allowing glutenin and gliadin to interact, effectively "shortening" the gluten strands. The resulting finished food is more flaky, crumbly, and tender, like a pie crust or croissant roll, rather than stretchy and springy like bread.

Mouthfeel: The name given in Sensory Science to the experience of how a food feels when it's in your mouth. For example, crunchy, smooth, and crumbly are all adjectives to describe different characteristics of *mouthfeel*.

Fat Freezing Point: Interestingly, the freezing point for many substances—at what temperature they become solid—is well above the freezing point of water. Stainless steel melts at about 2,750°F. At any temperature below that, it's frozen, or solid. Many fats will also freeze at temperatures higher than 32°F.

Similar to other materials, when fats freeze and become solid in cold temperatures, they become denser. The knowledge of this phenomenon helped me solve my first big production problem while I was working at Gorton's. The production facility was par-frying (which means to partially fry) a delicate tempura battered product. They followed that with blast freezing, conveying the product to the packaging equipment, and ultimately packaging it.

The recurring problem was that sometimes oil would leak out of the frozen product onto the conveying equipment, eventually building up enough to degrade the delicate exterior. At other times, the oil didn't leak out. I determined that the problem was a faulty stirring impeller (a long rotating bar shaped kind of like a propeller) in the frying oil storage tank, combined with the fact that the oil was a combination of various chain lengths. Some were solid at storage temperatures, while others weren't. Since the impeller wasn't moving, the solid material was sinking to the storage tank bottom, resulting in stratified (or separated) material that was drawn off and used at different times. The solution was fortunately simple: fix the impeller and turn it back on!

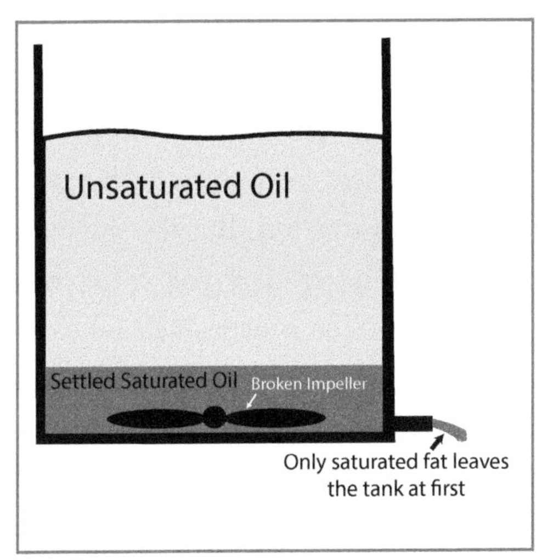

Settling fat with impeller[71]

Multiple natural sources of fat that are solid at room temperature exist. They include what are called tropical oils, such as coconut or palm oils. Beef fat, which is sometimes called *beef tallow*, is also mostly solid at room temperature.

STORAGE AND SHELF LIFE CONSIDERATIONS: RANCIDITY

Another big problem for food scientists when designing foods is preventing them from becoming rancid. The unpleasant aromas and flavors associated with oxygen reacting with fats are called *oxidative rancidity*. Almost all foods contain fat, even grains. *Oxidation* means for something to chemically react with oxygen. All fats will eventually oxidize, with the reaction causing products to be unpleasant to the taste, or even unhealthy.

As a result, a food scientist wants to control rancidity. It must be prevented for as long as possible, which should be at least the longest expected elapsed time from production to consumption. The choice of fat is an important part of this effort. Saturated fats are stable and don't oxidize easily, while unsaturated fats are more susceptible. The more unsaturated (number of double bonds, or kinks, in the fatty acid chain), the more susceptible a product is to oxidative rancidity. It's ironic that, according to our current understanding, polyunsaturated fats, which are most susceptible to becoming rancid, are also the healthiest fats to consume. You just need to consume them while they're fresh or protect them from becoming rancid. For example, fish oils are highly unsaturated (lots of bends or kinks) and are known to be better for you then beef fat, which is mostly saturated. But fish oils become rancid quickly, as you may have noticed if you have warmed up leftover fish. The "fishy" smell is partly due to oxidizing oils.

Sometimes you need to have the characteristics of a certain fat even though it's susceptible to becoming rancid. It might also happen that the needed shelf life is long enough that the danger of rancidity is still a predictable problem. In these cases, ingredients can be added that will delay fat oxidation by interfering with the pro-oxidants that cause it. This will slow down the process and extend the product's shelf life. Some are natural, while others are manufactured, and collectively they're known as *antioxidants*. The antioxidant that a food scientist uses is based on the consumer to whom the manufacturer would like to sell the product.

Packaging can be designed to limit the entry of oxygen, and it's also possible to flush products with a non-reactive gas. This prevents the initiation of oxidative rancidity until the package is opened. For example, many packaged nuts are flushed with nitrogen, which generally doesn't react at all with any product components. Once the product is opened, the nearly 100 percent nitrogen environment within the sealed container is replaced by atmospheric gases, including oxygen, and the process of oxidative rancidity begins!

Kitchen Field Trip: Go into your kitchen and find a bottle of oil, such as olive, corn, sunflower seed, etc., that's sitting in your cupboard at room temperature. Is it solid or liquid? Next, find a can of shortening, such as a container of Crisco® (a registered trademark of the J.M. Smucker Company), also at room temperature. Is it solid or liquid?

EXPERIMENTS: LET'S MAKE A MESS!

Fat Experiment #1: Mayonnaise Separation

Background

In this experiment we'll investigate a common food, mayonnaise, that's made mostly from fat, and learn about how it's made. We'll do this by destroying the mayonnaise.

Items Needed

1 tablespoon mayonnaise, any brand. (Fat-free, reduced fat, or other types of sandwich spreads will give different results than described in this experiment.)

- Small saucepan
- Spoon
- Procedures

1. Look at the mayonnaise container's Nutrition Facts and write down the ingredients.
2. Place approximately 1 tablespoon of mayonnaise in the small saucepan. This doesn't have to be exact.

3. Place the saucepan on a stove at medium heat, stirring constantly with a spoon until the white color disappears. Remove from heat and set aside. Describe your observations:

a. What happened to the mayonnaise?
b. Is there any discoloration?
c. What does it smell like?

4. Let it cool to room temperature. What does it look like now?

Fat Experiment #1: Mayonnaise Separation Discussion

Mayonnaise is an *emulsion*—a stable product that consists of two or more components that don't normally mix. In this case, water and oil don't mix together. An emulsifier is required to bridge the gap between the two.

Emulsifiers have a hydrophilic (water loving) and a hydrophobic (water hating) end on the same molecule. The hydrophilic end interacts with the water and the hydrophobic end interacts with the oil, stabilizing the emulsion. In mayonnaise, egg yolk is used since it contains several natural emulsifiers. The process of creating the emulsion must be done carefully or the emulsion will break.

Heating the emulsion is one way to deliberately break it. Once the white color went away, you should have ended up with oil in the pan, as well as some solids that may have browned. The solid material is a mixture of egg yolks, which contain protein, and other solids and sugar. These are also the source of any browning you may have seen.

If it's an emulsion of water and oil, where did the water go? If you were watching closely, you likely saw bubbling around the edges and some vapor escaping as you were heating the mayonnaise. Part of what you were seeing was the liquid water vapor being converted into a gas. The answer to "Where did the water go?" is that it's now in the air in your house.

Upon cooling, you should note that the oil remains liquid. It would be very difficult to make mayonnaise out of a fat that was solid at room temperature.

Emulsifier: Ingredients that contain naturally occurring emulsifiers are egg yolk, honey, mustard, and soy lecithin.

Making mayonnaise by hand.
https://www.youtube.com/watch?v=moz_zNPdbhI

Fat Experiment #2: Shortening Change of State

Background

In this experiment we'll see how shortening can change from a solid to a liquid and back again.

Items Needed

- 1 tablespoon vegetable shortening, such as Crisco, that's solid and white at room temperature
- Small saucepan

Procedures

1. Place approximately 1 tablespoon of shortening in the saucepan.
2. Heat on medium heat until it begins to melt. Stirring isn't needed. Record your observations.
3. Heat the sample until all the white color is gone. Remove from heat. Record your observations.

4. Let cool until fat is again room temperature throughout, recording your observations. It may take a couple of hours for the sample to return to room temperature. Record your observations.

Fat Experiment #2: Shortening Change of State Discussion

The shortening should have melted. Think of the room as the "freezer" for this type of fat. By placing it in the pan and heating it, you took it out of the freezer and, depending on the temperature it reached, perhaps even put it into an oven. Just as ice melts outside of a normal freezer, the shortening melted outside of its freezer and should have turned clear.

Upon returning completely to room temperature, the shortening will again "freeze" and become solid. This sometimes takes a while. It could even be overnight, depending on several factors, including the pan's cleanliness and the oil's purity.

WHAT DO YOU THINK?

Similar to this book's other chapters, we only touched on the wonderful intricacies of this section's subject: fats in food. I hope that the discussion and experiments have prompted you to explore a little more. If you'd like some assistance in that effort, please go to www.beakersandbricks.com and see what new offers we have to share.

JOURNALING IDEA

Write about your favorite food. Break it down according to its proximate analysis. Is it a healthy food, or something that should be consumed in moderation? Write about why it's your favorite food. Include descriptions of how it makes you and your mouth feel when the food is consumed. Think and write about how, as a food scientist, you'd try to remove all the fat or sugar from the food and still make it your favorite food. It's hard!

CHAPTER REVIEW

Write short essays in response to five of the following questions:

1. How might science continue to evolve regarding any of the subjects you learned about in this book?
2. How do your language classes, such as studying English or Spanish, help prepare you for a career in food science? Is this different than if you decided to pursue a career in medicine?
3. Doughnuts are fried in fat that's solid at room temperature. Why might this be important? Describe a doughnut fried in a fat that's liquid at room temperature.
4. Chocolate (mostly fat) that's considered premium melts at a temperature just below what's normal in your mouth. Less expensive chocolates will incorporate fats or other ingredients that raise the chocolate melting point. Describe what effect this may have in the mouth.

5. You may have heard that a particular food is a *good source* of protein, or some other component that's beneficial for the body. What does this mean? What's the definition of a "good source?" How might your new food company, ABC Foods, develop a food product that health conscious people might want to consume?
6. Continuing from the last question, think of three products ABC Foods can develop and describe some challenges associated with their development and formulation. How long would the shelf life be? What would the name be? What would the packaging look like?
7. What have you found most interesting about your introduction to food science? What have you found least interesting?
8. Did all the experiments you conducted work as expected? If not, why didn't they?
9. Evaluate the ingredients of real butter. Examine each ingredient. Does butter contain sugar? Protein?
10. How might your mathematics courses help you as a food scientist?

CONCLUSION… OR THE BEGINNING!

This workbook is an *overview* of food science, only lightly touching on the main components. If you find that you'd like to know more, please review the General Resources section at the end of the book, and perhaps even the included bonus chapter on education.

You may also like to visit www.beakersandbricks.com and see what other workbooks have been released.

I hope that you've enjoyed going through this workbook and acquiring some Edible Knowledge®!

CONTACT INFORMATION

Email: info@dalewcox.com

Address: Beakers & Bricks, LLC
PO Box 1014
Asheboro, North Carolina 27204
USA

WEBSITES

www.beakersandbricks.com
www.dalewcox.com

GENERAL RESOURCES

Bennion, Marion. *The Science of Food.* HarperCollins Publishers, NY: 1998.

Carey, Francis A., and Robert M. Giuliano. *Organic Chemistry.* McGraw-Hill Education, NY: 2016.

Cox, Dale W. *Introduction to Food Science: Water.* Beakers and Bricks, Asheboro, NC: 2019.

Fennema's Food Chemistry, 5th Ed. Damodaran, Srinivasan, and Kirk L. Parkin, eds. CRC Press, Boca Raton, FL: 2017.

Goates, J. Rex. *General Chemistry, Theory and Description.* Harcourt Brace Jovanovich, San Diego, CA: 1981.

Jay, James M., Martin J. Loessner, and David A. Golden. *Modern Food Microbiology, 7th Ed.* Springer Publishing Company, NY: 2006.

Pomeranz, Yeshajahu, and C.E. Meloan Pomeranz. *Food Analysis: Theory and Practice, 4th Ed.* Springer Publishing Company, NY: 2018.

Vaclavik, Vickie A., and Elizabeth W. Christian. *Essentials of Food Science, 4th Ed.* Springer Publishing Company, NY: 2013.

www.aaccnet.org The official website of The American Association of Cereal Chemists.

www.beakersandbricks.com The official website of the publisher of the Edible Knowledge® workbooks.

www.fda.gov/food The official website of the Food and Drug Administration.

www.IFT.org The official website of The Institute of Food Technologists. IFT holds an annual conference with a huge show floor that you can attend. This website will list the location.

BONUS SECTION!

EDUCATIONAL REQUIREMENTS

In Chapter 1, I briefly discussed the education required in order to become a professional food scientist. Here we'll look at it more closely, using two schools as examples.

CURRICULUM

A degree in food science requires the study of most of the sciences. In addition, many schools have different specialty tracks depending on your area of interest. North Carolina State University has technical tracks for students wanting to pursue R&D, and management tracks for those wanting to move in that direction. On the following pages, the basic curriculum for North Carolina State is listed. This information was taken directly from their website. There are approximately 120 credit hours total to be completed for graduation.

Fall Semester	Credit	Spring Semester	Credit
ALS 103 Transitions & Diversity	1	CH 101 Chemistry – A Molecular Science	3
BIO 183 Intro Bio: Cellular & Molecular	4	CH 102 General Chemistry Lab	1
ENG 101 Acad Writing & Research	4	FS 201 Introduction to Food Science	3
MA 107 Precalculus I	3	MA 131 Analytic Geom & Calc A	3
GEP Social Science	3	GEP Humanities	3
HES_*** Health & Exercise Studies	1	HES_*** Health & Exercise Studies	1
SOPHOMORE YEAR			
Fall Semester	**Credit**	**Spring Semester**	**Credit**
CH 221 Organic Chemistry I	3	CH 223 Organic Chemistry II	3
CH 222 Organic Chemistry I Laboratory	1	CH 224 Organic Chemistry II Lab	1
FS 290 Careers Food & Bioprocess Sci	1	FS 231 Princ. Food & Biopro. Engin.	4
MA 231 Analytic Geom & Calculus B	3	PY 212 College Physics II	4
MA 132 Computational Math	1	COM 110 Public Speaking **or**	3
PY 211 College Physics	4	COM 112 Interpersonal Comm	
GEP Interdisciplinary Persp	3		
JUNIOR YEAR			
Fall Semester	**Credit**	**Spring Semester**	**Credit**
CH 201 Chemistry– Quantitative	4	BCH 351 Elementary Biochemistry	4
CH 202 Quantitative Chem Lab	1	FS 403 Analytical Tech Food & Bioprocess Sci	4
FS 402 Chemistry Food, Bioprocess Materials	4	FS 405 Food Microbiology	3
MB 351 General Microbiology	3	FS 406 Food Micrbiology Lab	1
MB 352 General Microbiology Lab	1	GEP Additional Breadth	3
GEP Interdisciplinary Persp	2-3		
SENIOR YEAR			
Fall Semester	**Credit**	**Spring Semester**	**Credit**
FS 421 Food Preservation	3	FS 475 Probs, Design Food, Bioprocess Sci	3
NTR 301 Intro to Human Nutrition	3	Food Science Elective	3
Food Science Elective	3	Free/Minor Elective	3
GEP Humanities	3	Free/Minor Elective	3
ST 311 Introduction to Statistics	3	GEP Social Science	3

NCSU Curriculum[72]

North Carolina State University Bachelor of Food Science Requirements

The curriculum covers all major science disciplines including chemistry, biology, microbiology, physics, engineering, and materials science.

GOALS

Many people use a background in food science as a stepping-stone to something else. For example, to become a medical doctor, lawyer, or dentist, a four-year undergraduate degree is required. It's wise to choose a degree in which you can make money in case your original plans don't work out the way you intend. Medical school entrance exams require a thorough understanding of the sciences, so food science is a good major for that field.

Depending on the curriculum at the individual school and your post-graduate employment or education goals, you could easily obtain a minor in chemistry, or some other science, by completing a small number of additional classes.

Track: Sequence of courses or classes. For example, Chemistry 106 and 107.

WHAT DO YOU THINK?

If you're approaching the time to go college, or are interested in thinking ahead, perhaps the following questions will help you determine if food science is something to consider.

1. **What college or university do you want to attend?**

2. **Access that university's website and see if they offer a food science degree. If they don't, find at least two universities that offer the major. Evaluate them and their programs based on the following information:**

   a. **What are the areas of specialization?**
   b. **What are the requirements for university admittance?**
   c. **How much does a typical four-year education cost?**

3. **If you have a local university with a food science department, contact them to schedule a tour of their facilities. You might want to talk to the food science department head while you're there.**
4. **If there's no local university, you can call your chosen university and with a food science professor. You should ask the professor the following questions, taking careful notes:**

    a. **How did you become interested in food science?**
    b. **Can you tell me your career path—how did you get to where you are?**
    c. **How many food science majors do you have?**
    d. **What percentage of your graduates are placed in jobs prior to leaving school? Is there a student in the food science club that I can speak with? You can explain that you want to ask them similar questions.**

JOURNALING IDEA

Describe your ideal job or career path. Date this and put it somewhere so you can find it in 20 years. You'll find it very interesting to see how things may have changed (or not!) as your career and life progressed.

PICTURE AND ILLUSTRATION ATTRIBUTIONS

1. Picture of spoonful of cereal: Bauer, Scott. USDA ARS. A spoon containing breakfast cereal flakes, part of a strawberry, and milk is held in midair against a blue background, August 2000, public domain. United States Department of Agriculture, Agricultural Research Service. (https://commons.wikimedia.org/wiki/File:Spoonful_of_cereal.jpg).
2. Picture of a stack of books: Dale W. Cox
3. Graduation cap illustration: OOfalkonOO [CC BY-SA 4.0 (https://creativecommons.org/licenses/by-sa/4.0)], 27 March 2016. https://upload.wikimedia.org/wikipedia/commons/f/f4/Graduation-cap-239875.jpg. Licensed under the Creative Commons Attribution-Share Alike 4.0 International license. Accessed 5 March 2019.
4. Picture of a taste test in progress: Porter, Crispin; Bogusky; Filmed on location in the corporate offices of The World of Coca-Cola in Atlanta, Georgia, 8 September 2011 (https://commons.wikimedia.org/wiki/File:Blind_taste_test.jpg). Licensed under the Creative Commons Attribution 3.0 Unported license. https://creativecommons.org/licenses/by/3.0/deed.en. Accessed 5 March 2019.
5. Engineering illustration: Anitaslbn. Bahasa Indonesia: logo civiltekno.blogspot.co.id, 2 December 2017 (https://commons.wikimedia.org/wiki/File:Ct-logo.gif). Licensed under the Creative Commons Attribution-Share Alike 4.0 International license. https://creativecommons.org/licenses/by-sa/4.0/deed.en. Accessed 5 March 2019.
6. Picture of a digital scale: TBBSC. High Precision Digital Scale, 10 August 2015 (https://commons.wikimedia.org/wiki/File:TBBSC_Digital_Scale.jpg). Licensed under the Creative Commons Attribution-Share Alike 4.0 International license. https://creativecommons.org/licenses/by-sa/4.0/deed.en. Accessed 5 March 2019.
7. Scale illustration: Work of the United States Government, Scale of justice, public domain (https://commons.wikimedia.org/wiki/File:Scale_of_justice.png).
8. Flask illustration: Wardrop, Matthew. Chemistry Flask, from the Open Clip Art Library, public domain (https://commons.wikimedia.org/wiki/File:Chemistry_flask_matthew_02.svg). Accessed 5 March 2019.

9 General Mills logo picture: Dale W. Cox

10 Gorton's logo picture: Dale W. Cox

11 Post logo picture: Dale W. Cox

12 "Brag Box" picture: Dale W. Cox

13 Planter's logo picture: Dale W. Cox

14 Kellogg company logo picture: Dale W. Cox

15 Malt-O-Meal logo picture: Dale W. Cox

16 Simple hydrogen atom illustration: Dale W. Cox

17 Periodic table of the elements: LeVan-Han, Cédric. Periodic table of the elements, 2008,[CC BY-SA 3.0 (https://creativecommons.org/licenses/by-sa/3.0)]. Accessed 23 April 2019.

18 Hydrogen Bonds: Manas, Michal. 3D Model Hydrogen Bonds in Water (https://commons.wikimedia.org/wiki/File:3D_model_hydrogen_bonds_in_water.jpg). Hydrogen and oxygen charges removed by Dale W. Cox for simplicity. Licensed under the Creative Commons Attribution-Share Alike 3.0 Unported License. https://creativecommons.org/licenses/by-sa/3.0/deed.en. Accessed 04 February 2019.

19 Bond Strength Illustration: Dale W. Cox

20 Fahrenheit vs. Celsius illustration: Dale W. Cox

21 Fahrenheit sign illustration: Dale W. Cox

22 Celsius sign illustration: Dale W. Cox

23 OpenStax College, Anatomy & Physiology, Connexions Website (http://cnx.org/content/col11496/1.6/; https://commons.wikimedia.org/wiki/File:2713_pH_Scale-01.jpg). 19 June 2013. Licensed under the Creative Commons Attribution 3.0 Unported license. https://creativecommons.org/licenses/by/3.0/deed.en.

24 Picture of Hostess Twinkies: Moore, Larry D. Twinkies (Hostess Twinkies is a trademark of Hostess Brands), 28 February 2005 (https://commons.wikimedia.org/wiki/File:Hostess_twinkies_tweaked.jpg). Licensed under the GNU Free Documentation License and the Creative Commons Attribution-Share Alike 3.0 Unported license. https://en.wikipedia.org/wiki/en:GNU_Free_Documentation_License. https://creativecommons.org/licenses/by-sa/3.0/deed.en. Accessed 5 March 2019.

25 Picture of carrots: Wouters, Frank. Antwerpen, Belgium. Carrots Julienne, 21 June 2005 (https://commons.wikimedia.org/wiki/File:-Carrots_Julienne.jpg). Licensed under the Creative Commons Attribution 2.o Generic license. https://creativecommons.org/licenses/by/2.0/deed.en. Accessed 5 march 2019.

26 Picture of popcorn: Munro, Catherine. Popcorn in bowls, AV-8902-3817, National Institutes of Health, public domain (https://commons.wikimedia.org/wiki/File:Popcorn_NIH.jpg).

27 Picture of breakfast cereal: Miami U. Libraries – Digital Collectionss. Shredded Wheat with Strawberries, circa 1900, public domain (https://commons.wikimedia.org/wiki/File:Shredded_Wheat_(3092754943).jpg).

26 Shredded wheat ingredients picture: Dale W. Cox

287 Picture of green wheat: Wikipedia User "H20." Wheat in the Hula Valley, 2007, public domain. Modifications by Carol Spears. https://commons.wikimedia.org/wiki/File:Wheat-haHula-ISRAEL2.JPG.

29 Picture of dry wheat: Ximenex. Campo de trigo, Valle de Ablés, Ávila, Spain. 30 June 2003. Licensed under the Creative Commons Attribution-Share Alike 3.0 Unported License. https://creativecommons.org/licenses/by-sa/3.0/deed.en.

30 Picture of wheat harvest: Picture Reference: Wikipedia User "Michael Gabler." The wheat harvest near Eldagsen in Germany. Public Domain. https://commons.wikimedia.org/wiki/File:Unload_wheat_by_the_combine_Claas_Lexion_584.jpg.

31 Drawing of a sugar beet: Masclef, Amedee. Atlas des plantes de France, 1891. Public Domain.

32 "Growing & Processing Sugarbeets." Michigan Sugar Company, © 2015 Michigan Sugar. http://www.michigansugar.com/growing-production/from-seed-to-shelf/. Accessed 02 February 2019.

33 Counter-current flow illustration: Dale W. Cox

34 Water picture: Dale W. Cox

35 Picture of ash: Dale W. Cox

36 Picture of butter: Dale W. Cox

37 Nutrition Facts, old and new. Changes to the Nutrition Facts Label , last updated 8 February 2019. U.S. Food and Drug Administration. https://www.fda.gov/downloads/Food/GuidanceRegulation/GuidanceDocumentsRegulatoryInformation/LabelingNutrition/UCM501646.pdf. Accessed23 April 2019

38 Picture of Kellogg's Raisin Bran side panel: Dale W. Cox

39 Proximate analysis of raisin bran illustration: Dale W. Cox

40 Proximate analysis of raisin bran illustration, showing the math: Dale W. Cox

41 Glass of water picture: Dale W. Cox

42 Water molecule illustration: Sakurambo [Public domain], August 2008. https://upload.wikimedia.org/wikipedia/commons/0/07/Water_molecule.svg. The text "hydrogen" and "oxygen" added by Dale W. Cox.

43 Freshly sliced bread illustration: Dale W. Cox

44 Partially dried bread illustration: Dale W. Cox

45 Equilibrated bread illustration: Dale W. Cox

46 Picture of iceberg: Walk, Ansgar. Iceberg near north-eastern coast of Baffin Island, 7 August 1997 (https://commons.wikimedia.org/wiki/File:Iceberg_1_1997_08_07.jpg). Licensed under the Creative Commons Attribution-Share Alike 2.5 Generic License. https://creativecommons.org/licenses/by-sa/2.5/deed.en. Accessed 5 March 2019.

47 Illustration of hydrogen bonds in water: Manas, 3D Model Hydrogen Bonds in Water.

48 Carbohydrates picture: Dale W. Cox

49 Picture of sugar crystals: Scrubjay. Sugar crystals from Dunkin Donuts sugar packet, 19 March 2018 (https://commons.wikimedia.org/wiki/File:Sugar_Crystals.jpg). Licensed under the Creative Commons Attribution-Share Alike 4.0 International license. https://creativecommons.org/licenses/by-sa/4.0/deed.en. Accessed 5 March 2019.

50 Glucose, Fischer projection illustration: Dale W. Cox

51 Fructose, Fischer projection illustration: Dale W. Cox

52 Glucose, Haworth Projection: Cacycle [Public domain], 4 September 2008, https://upload.wikimedia.org/wikipedia/commons/5/5b/Beta-D-glucose_Haworth_formula.png.

53 Picture of loaf of bread: Monniaux, David (2004), with modifications by User Hohum (2008). A loaf of French Bread, (https://commons.wikimedia.org/wiki/File:French_bread_DSC09293.jpg). Licensed under the Creative Commons Attribution-Share Alike 1.0 Generic

license. https://creativecommons.org/licenses/by-sa/1.0/deed.en. Accessed 5 March 2019.

54 Amylose illustration: Dale W. Cox

55 Amylopectin illustration: Dale W. Cox

56 Raw granules illustration: Dale W. Cox

57,58 Partially gelatinized granules illustration: Dale W. Cox

59 Fully gelatinized granules illustration: Dale W. Cox

60 Broken granules illustration: Dale W. Cox

61 Protein structure illustration: Dale W. Cox

62 Enzyme + substrate illustration: Dale W. Cox

63 Picture of bread preparation: Wikipedia User "BotMultichillT." US Navy ID 050102-N-5837R-008, 2005, public domain. https://commons.wikimedia.org/wiki/File:US Navy 050102-N-5837R-008 Culinary Specialist 3rd Class Joshua Savoy prepares bread in the bakery aboard the Nimitz-class aircraft carrier USS Abraham Lincoln (CVN 72).jpg#filelinks

64 Picture of cow: Wikipedia User "Geograph-Bot." Evelyn Simak/ Aberdeen Angus Bull/ CC BY-SA 2.0, 2007. https://upload.wikimedia.org/wikipedia/commons/9/96/Aberdeen Angus bull - geograph.org.uk - 546924.jpg

65 Enzyme + disruptor illustration: Dale W. Cox

66 Apple browning: Dale W. Cox

67 Hot dogs and pepperoni picture: Dale W. Cox

68 Butter picture: Dale W. Cox

69 Glycerol illustration: Dale W. Cox

70 Triglyceride structure illustration: Dale W. Cox

71 Settling fat with impeller illustration: Dale W. Cox

72 North Carolina State University Department of Food, Bioprocessing, & Nutrition Sciences. https://fbns.ncsu.edu/academic-programs/bachelors-degrees/bs-food-science/. Accessed 04 February 2019.

GLOSSARY

Note: These definitions are in the context of food science and at the scientific depth of the *Edible Knowledge®* workbooks.

Acidity: More H+ ions than pure water; a pH below 7.0, with increasing strength of acidity as the pH lowers. In food products, acidity is often perceived as tasting sour.

Aldehyde: A common organic compound involving a carbon double bonded to an oxygen atom, situated at the end of the molecule. It is important in many reactions and is frequently the source of characteristic aroma and flavor in many foods.

Amino Acids: The building blocks of protein.

Amylopectin: A long, branched carbohydrate; one of the components of starch.

Amylose: A long, straight-chained carbohydrate; one of the components of starch.

Antioxidant: A substance that prevents or slows fat oxidation.

Ash: What's left from food after incineration has removed all organic material; contains mostly minerals.

a_w: See Water Activity.

Basicity: More OH- ions than pure water; a pH above 7.0. In food products, basicity is often perceived as tasting bitter.

Batch Cooker: A machine that's loaded with food, heats and cooks it, and then discharges it before the process repeats. A simple example is cooking something in a pot on your stove.

Beef Tallow: A natural fat from beef; mostly solid at room temperature.

Berry: When used to describe grain, is usually an individual grain of wheat.

Biology: The study of life.

Bipolar: Having two oppositely charged ends on the same molecule.

Browning Reactions: A general term used to refer to both enzymatic and non-enzymatic reactions that will product brown colored compounds. See also *Enzymatic Browning* and *Non-Enzymatic Browning*.

Calcium Ascorbate: A natural compound that can prevent polyphenol oxidase enzymes from catalyzing (speeding up) enzymatic browning reactions.

Carbohydrate: The generic term for carbon-containing molecules, such as sugars and starches.

Catalyst: Something that will speed up a reaction, such as enzymes.

Catalyze: To speed up a reaction that would have occurred in nature given sufficient time. Enzymes catalyze reactions by bringing reactants together in such a way that they react faster.

Celiac Disease: A disorder in humans where foods containing gluten aren't properly metabolized, resulting intestinal distress and, in some cases, damage to the lining of the intestines.

Celsius: Part of the metric system of measurements, specifically for temperature.

Centrifuge: A machine designed to separate by density items in a fluid. For example, separating cream from the rest of milk.

Charge: Describing the electrical nature of an object, which can be positive or negative and anywhere in between, including neutral.

Chemistry: The study of how chemical elements react with one another, including the formalization of rules based on observations of chemical reactions.

Collagen: A connective tissue found in muscles, skin, bones, and tendons.

Composite Analysis: See *Proximate Analysis.*

Compositional Analysis: See *Proximate Analysis.*

Compound Sugar: Two or more simple sugars chemically bonded together. For example, glucose and fructose bonded together make sucrose, or common table sugar. See also *Disaccharide.*

Connective Tissue: A general term for protein containing structures of the body such as skin, muscle, ligaments, and tendons.

Consumer Testing: See *Sensory Science.*

Continuous Cooker: A machine that heats and cooks food material as the food is transported continuously through the equipment. An example is a blancher used to cook wheat for shredded wheat.

Convection Oven: An oven that uses air movement as well as heat to bake a food material, accelerating the process and making surface effects more readily possible, such as toasting.

Convenience Food: A general term used to describe packaged food, or food that's convenient to use. See also *Processed Food.*

Cossette: In this workbook's example, a cossette is a thin slice of sugar beet that is designed to maximize its surface area and promote diffusion of beet juice.

Counter-Current: A process where a fresh fluid being treated and the fluid providing the treatment run in opposite directions. In a cleaning example using water, such as for sugar beets, this ensures that the cleanest water is being used on the cleanest beets.

Covalent Bonding: The strongest type of chemical bond.

Crystal: In food, this refers to the solid, most stable form of a pure substance. For example, salt or sugar crystals.

Denature: A process by which the structure of protein is disrupted, including the quaternary, tertiary, and secondary structures, resulting in the loss of protein functionality.

Density: The measure of mass divided by volume; more mass in the same space equals increased density.

Dew Point: The temperature at which dew forms, or in other words, when the air is saturated with water at a constant pressure.

Die, Extruder: The orifice through which material is forced during extrusion. This orifice is shaped to achieve the desired product shape. In the case of pasta, a different shaped die can product elbow noodles and spaghetti.

Diglyceride: A glycerol molecule with two fatty acids attached.

Disaccharide: Two simple sugars chemically bonded together. For example, glucose and fructose bonded together make sucrose, or common table sugar. See also *Compound Sugar*.

Disparity: The difference between more than one thing.

Dissolve: The action when a solid substance becomes liquid in a solvent.

DNA: Deoxyribonucleic acid, the natural blueprint from which all living things are constructed.

Double-Bonded: A chemical bond in which more electrons are shared than in a single bond.

Electron Orbitals: The area described by the paths that electrons take while moving around an atomic nucleus.

Emulsifier: A component used to stabilize emulsions; usually consisting of a hydrophobic and a hydrophilic component on the same molecule.

Emulsion: A stable mixture of two or more components that normally don't mix; usually stabilized by an emulsifier.

Engineering: The application of knowledge, such as mathematics, science, and observed data, to the physical world by designing and making machines and processes that accomplish tasks.

Enzymatic Browning: Browning of foods through enzymatic action, such as the browning of cut apple slices.

Enzyme: Naturally occurring proteins that serve to catalyze chemical reactions.

Equilibrate: To become equal. In the case of food, the term is often used to describe exchange of water between mixed ingredients that don't have the same aw.

Equilibrium Relative Humidity: The relative humidity of the air in an enclosed container once the product in the container equilibrates with the environment and no more exchange of moisture is taking place.

ERH: See *Equilibrium Relative Humidity*.

Extruder: Usually a screw in a shaft with an attached faceplate and die assembly. There are many types, including extruders that cook the food while extruding it (corn curl snacks), and also forming extruders (pasta).

Extrusion: Using an extruder to force something through a hole; in the food industry, this is called a "die."

Fahrenheit: Part of the English system of measurements, specifically for temperature.

Fatty Acid: The individual chains that, when combined on a glycerol molecule, make up fats.

Fischer Projection: A way to represent carbohydrates in a single, straight line.

Food Modernization Act (2010): A legislative act that changed requirements for nutrition labeling and gave food company oversight authority to government agencies.

Food Preservation: A general term describing either processes or ingredients used to help food maintain flavor and texture and remain safe to consume for longer than it otherwise would.

Food Processing: The application of food science to the selection, preservation, processing, packaging, distribution, and use of safe food.

Food Science: The study of the physical, biological, and chemical makeup of food and the concepts underlying food processing.

Food Technology: See *Food Processing.*

Fortification: When vitamins and/or minerals are purposefully added to a food to provide additional nutrition.

Freeze and Thaw Cycle: See *Frost-Free Freezer.* A cycle that is particularly important to food scientists working on frozen food products, as they must design foods to withstand this abusive treatment.

Frost-Free Freezer: A machine designed to keep things frozen but also prevent the buildup of ice, or frost, on the inside of the freezer as the result of air moisture condensing and freezing on the inside of the freezer when the door is opened. This is accomplished by cycling through freeze and thaw periods, which allow any exterior solid water to melt and be transported out of the freezer.

Fructose: A monosaccharide that's sweeter than glucose; also known as levulose.

Fulfillment Test: A consumer test designed to see if a newly developed product fulfills what consumers expect when they read the concept.

Fully Hydrogenated: The result of an artificial process where hydrogen is fully reacted with fats and removes all points of unsaturation, making the fatty acid chains straight.

Function: The purpose of an ingredient in a food product.

Functionality: See *Function.*

Gel: A generic term for a semi-solid that could be made from either protein, such as gelatin, or carbohydrate, such as jelly or jam.

Gelatin: A purified protein made from collagen; used for many foods and other purposes to provide strength and structure, among other things.

Gelatinization: When starch granules swell and take up water while being cooked.

Gliadin: See *Gluten.*

Gluten: A protein combination formed by the interaction of glutenin and gliadin proteins—primarily found in wheat—that's stretchy and strong, producing the structure found in baked goods such as breads.

Glutenin: See *Gluten.*

Glycerol: A three-carbon chain that's the backbone of fats.

Grams, Kilograms, Milligrams: Part of the metric system of measurements, specifically for measuring mass.

Granule, Starch: The compact forms in which starch is naturally laid down in natural products.

Haworth Projection: A way to represent carbohydrate ring structures in two dimensions.

Hydration: The addition of water.

Hydrogen Bonds: A type of chemical bond that exists between the hydrogen molecules in water; important for understanding how water behaves that way that it does.

Hydrogenated: See *Hydrogenation.*

Hydrogenation: An artificial process where hydrogen is reacted with fats to remove points of unsaturation, making the fatty acid chains straighter than they were before. The result is a fat that is more solid at room temperature than it otherwise would be.

Hydrophilic: Water "loving"; an example is alcohol.

Hydrophobic: Water "hating"; an example is oil.

Hygroscopic: The affinity or attraction a substance has for water.

Hygroscopicity: See *Hygroscopic.*

Incineration: An analytical laboratory action for determining the ash content of food by burning it in a controlled and complete fashion.

Institute of Food Technologists (IFT): The Institute

of Food Technologists. From www.ift.org/about-us: "Since 1939, IFT has been advancing the science of food and its application across the global food system by creating a dynamic forum where individuals from more than 90 countries can collaborate, learn, and grow, transforming scientific knowledge into innovative solutions for the benefit of people around the world."

Ionic Bonding: A type of chemical bonding that's intermediate between hydrogen and covalent bonds in strength.

Ketone: A common organic compound; in food, often a product of fat oxidation.

Kjeldahl Method: An analytical laboratory method for determining the amount of protein in a food.

Lachrymal Gland: The organ in the human body responsible for producing tears.

Levulose: See Fructose.

Lexicon: A general term describing the combined vocabulary related to a topic or discipline.

Ligament: The connective tissue that connects bone to bone, such as in joints.

Lipids: Fats.

Maillard Browning: See Non-Enzymatic Browning.

Metabolize: To incorporate into the body. Food is metabolized by humans to become part of the body.

Meters, Kilometers, Millimeters: Part of the metric system of measurements, specifically for measuring length.

Microbiology: The study of bacteria, fungi, yeasts, and molds and their life cycles. In food, it's the study of the application of these life forms to making and preserving food.

Modified Starch: Starch that has been altered from its natural form to achieve a desired functionality, such as freeze/thaw stability.

Molality: The moles of solute divided by the mass of the solvent. See also Solvent and Solute.

Molarity: The moles of solute per liter of solution. See also Solvent and Solute.

Mole: 6.022 x 1023 of any molecule. See also *Molecular Weight.*

Molecular Weight: The weight of a mole of any molecule, including the weight of protons, neutrons, and electrons. See also *Mole.*

Molecule: The smallest amount of a substance that can't be split without resulting in something else. For example, two hydrogen atoms and one oxygen atom make one molecule of water. If you split the hydrogen off, you have something other than water.

Monoglyceride: A glycerol molecule with one fatty acid attached.

Monosaccharide: A carbohydrate consisting of one sugar, such as glucose or fructose.

Monosaccharide: One simple sugar, such as glucose.

Mouthfeel: The overall textural experience as sensed by the mouth while consuming a food.

Negative Charge: See *Charge.*

Non-Enzymatic Browning: The browning of foods accomplished by using protein and sugar.

Nutrition Labeling and Education Act of 1990: A sweeping legislative act that harmonized nutrition labels applied to packaged food products, among other things.

Organic: Having to do with life, such as carbohydrates and proteins.

Organoleptic: Describing the way something tastes and feels while being consumed using all the senses—including taste, texture, and smell.

Oxidative Rancidity: The results of fat oxidation reactions that result in off-flavors and aromas.

Partial Pressure: The pressure of a gas in an enclosed volume at a specific temperature.

Periodic Table: A table organizing all of the known elements according to observed characteristics, including atomic weight.

pH: A logarithmic scale used to describe the level of acid and base.

Polyphenol Oxidase: A family of enzymes responsible for enzymatic browning.

Polysaccharide: More than two sugars chemically bound together. For example, many molecules of glucose bound together make starch, one of the most common polysaccharides.

Poly-Unsaturated Fatty Acid: A fatty acid that contains more than one point of unsaturation, resulting in a chain that isn't straight.

Porkskin: The skin of a pig.

Positive Charge: See *Charge*.

Precipitate: The solid product of a chemical reaction.

Process Automation: The acts of designing a process that can run with minimal human intervention.

Processed Food: A general term to describe food products manufactured outside of the home. See also Convenience Food.

Protein: Nitrogen-containing compounds essential life that are made up of amino acids.

Protein, Primary Structure: The most basic form of protein, consisting of amino acids and the order in which they're linked.

Protein, Quaternary Structure: The complete three-dimensional structure of a protein complex consisting of more than one protein.

Protein, Secondary Structure: The folding and twisting of portions of a protein primary structure.

Protein, Tertiary Structure: The complete three-dimensional folding and twisting, and related interactions, within a single protein.

Proximate Analysis: Determining analytically the quantity of protein, carbohydrate, fat, water, and ash that are present in a food product.

Quality Assurance: Sometimes used as synonymous with Quality Control. Quality Assurance encompasses the actions of creating the processes and procedures that allow for Quality Control.

Quality Control: Processes and procedures designed to ensure the quality of food products is within specified parameters.

Rancidity: A fat destruction process that results in flavors and aromas that are objectionable.

Relative Humidity: The amount of water in the air relative to the environmental temperature and pressure. It is defined as the partial pressure of water vapor divided by the equilibrium vapor pressure of water at the same temperature.

Research and Development (R&D): The process of investigating and developing new products, processes, and documenting the associated new knowledge.

Retrograde: The process describing starch retrogradation where gelatinized starch molecules will tend to realign themselves and form interactions, resulting in toughening of the finished product. Stale bread demonstrates an example.

RH: See *Relative Humidity*.

Saturated Fat: A fully hydrogenated fat that contains straight-chained fatty acids; will have solidification temperatures higher than unsaturated fats.

Sensory Science: The science of procedures associated with consumers liking and preference.

Shelf-Life: The length of time a prepared food product will, under proper storage conditions, retain most optimal qualities, including texture, flavor, and appearance.

Shortening: A fat that's solid at room temperature; used to make baked goods more tender by interfering with gluten formation.

Shredding: A process for converting cooked grain into strands that are then used to create products with different shapes and number of layers of these strands.

Simple Sugar: A monosaccharide, such as glucose.

Single-Bonded: A chemical bond consisting of sharing two electrons.

Solute: The component of a solution that dissolves in the solvent. For example, sugar is the solute when it dissolves in water.

Solution: A liquid consisting of a solute or solutes and a solvent.

Solvent: The component of a solution in which the solute dissolves. For example, water is the solvent when dissolving sugar.

Starch: A long-chained carbohydrate composed of amylose and amylopectin that makes up the bulk of energy reserves in grains and legumes.

Starch Retrogradation: See *Retrograde.*

Substrate: The component that combines with an enzyme to produce a new product.

Tempered: Letting treated material rest and equilibrate.

Tendon: Connective tissue that connects a muscle to bone.

Texture: A food's physical characteristics, such as soft, hard, brittle, crunchy, etc.

Titration: An analytical laboratory method for taking a chemical reaction to completion in an objective, measurable, and predictive way that lets the quantity of the reactants be calculated.

Trans Fats: These occur in nature, but mostly are the byproducts of artificial hydrogenation processes that are not complete, resulting in an unsaturation point that's at an unnatural angle and mostly straight fatty acids.

Tropical Oil: Fats derived from tropical sources, such as palm kernel or coconut oil.

Unsaturated Fat: A fat that contains points of unsaturation, and therefore isn't straight-chained. Unsaturated fats have lower solidification temperatures than saturated fats.

Water Activity: Describes the relative availability of water in a food product, and ranges between 0 and 1.0. It is the partial pressure of water over a food divided by the partial pressure of water over pure water at the same temperature.

Wheat Berry: An individual grain of wheat.

CPSIA information can be obtained
at www.ICGtesting.com
Printed in the USA
BVHW051545090222
628343BV00004B/15